COMMON KNOWLEDGE

COMMON KNOWLEDGE:

THE CHALLENGE OF TRANSDISCIPLINARITY

Moira Cockell, Jérôme Billotte, Frédéric Darbellay,
Francis Waldvogel, Editors

EPFL Press
A Swiss academic publisher distributed by CRC Press

CRC Press
Taylor & Francis Group

Taylor and Francis Group, LLC
6000 Broken Sound Parkway, NW, Suite 300,
Boca Raton, FL 33487

Distribution and Customer Service
orders@crcpress.com

www.crcpress.com

Library of Congress Cataloging-in-Publication Data
A catalog record for this book is available from the Library of Congress.

EPFL Press

is an imprint owned by Presses polytechniques et universitaires romandes, a Swiss academic publishing company whose main purpose is to publish the teaching and research works of the Ecole polytechnique fédérale de Lausanne (EPFL) and other universities and institutions of higher learning.

Presses polytechniques et universitaires romandes
EPFL – Rolex Learning Center
Post office box 119
CH-1015 Lausanne, Switzerland
E-mail: ppur@epfl.ch
Phone: 021/693 21 30
Fax: 021/693 40 27

www.epflpress.org

ISBN 978-2-940222-32-2 (EPFL Press)
ISBN 978-1-4398-6331-2 (CRC Press)

Printed in France by Présence Graphique – N° d'imprimeur: 111037429

Table of Contents

Foreword

by Edward O. Wilson[1]

The World Knowledge Dialogue, is an important effort to seek unity among the great branches of learning, and thereby render knowledge more relevant to the understanding of the real world and the human condition within it.

Since the eighteenth century the great branches of learning have been classified into the natural sciences, the social sciences, and humanities. Today we have the choice between, on the one hand, trying to make the great branches of learning consilient – that is, coherent and interconnected by cause-and-effect explanation – or, on the other hand, *not* trying to make them consilient. Surely universal consilience is worth a serious try. After all, the brain, mind, and culture are composed of material entities and processes; they do not exist in an astral plane that floats above and outside the tangible world.

The most useful term to capture the unity of knowledge is surely *consilience*. It means the interlocking of cause-and-effect explanations across different disciplines, as for example between physics and chemistry, chemistry and biology, and, more controversially, biology and the social sciences. The word consilience was introduced in 1840 by William Whewell, the founder of the modern philosophy of science. It is more serviceable than the words coherence or interconnectedness, because its rarity of usage since 1840 has preserved its original meaning, whereas coherence and interconnectedness have acquired many meanings scattered among the different disciplines.

Consilience, defined then as cause-and-effect explanation across the disciplines, has plenty of credibility. It is the mother's milk of the natural sciences. Its material understanding of how the world works and its technological spin-off are the foundation of modern civilization. The time has come, I believe, to consider more seriously its relevance to the social sciences and humanities. I will grant immediately that belief in the possibility of

1 Edward O. Wilson is Pellegrino University Research Professor Emeritus in Entomology for the Department of Organismic and Evolutionary Biology at Harvard University.

consilience beyond the natural sciences and across to the other great branches of learning is not the same as science, at least not yet. It is a metaphysical world view, and a minority one at that, shared by only a few scientists and philosophers. Its best support is little more than an extrapolation of the consistent past success of the natural sciences. Its strongest appeal is in the prospect of intellectual adventure and, given even modest success, the value of understanding the human condition with a higher degree of certainty.

Why is this conjunction among the great branches of learning important? Because it offers the prospect of characterizing human nature with greater objectivity and precision, an exactitude that is the key to human self-understanding. The intuitive grasp of human nature has been the substance of the creative arts, the underpinning of the social sciences, and a beckoning mystery to the natural sciences. To grasp human nature objectively, to explore it to its depths scientifically, and to grasp its ramifications, would be to approach if not attain the grail of scholarship, and to fulfill the dreams of the Enlightenment.

Now, rather than let the matter hang in the air thus rhetorically, I want to suggest a preliminary definition of human nature, and then illustrate it with an example. Human nature is not the genes, which prescribe it. It is not the cultural universals, such as the incest taboos and rites of passage, that are the products of human nature. Rather, human nature is the collectivity of the epigenetic rules, the inherited regularities of mental development. These rules are the genetic biases in the way our senses perceive the world, the symbolic coding by which we represent the world, the options we open to ourselves, and the responses we find easiest and most rewarding to make. In ways that are beginning to come into focus at the physiological and even a few cases the genetic level, the epigenetic rules alter the way we see and linguistically classify color. They cause us to evaluate the aesthetics of artistic design according to elementary abstract shapes and the degree of complexity. They lead us differentially to acquire fears and phobias concerning dangers in the environment (as from snakes and heights), to communicate with certain facial expressions and forms of body language, to bond with infants, to bond conjugally, and so on across a wide range of categories in behavior and thought. Most are evidently very ancient, dating back millions of years in mammalian ancestry. Others, like the stages of linguistic development, are uniquely human and probably only hundreds of thousands of years old.

As an example of epigenetic rules, consider the instinct to avoid incest. Its key element is the Westermarck effect, named after Edward Westermarck, the Finnish anthropologist who discovered it a century ago. When two people live in close domestic proximity during the first 30 months in the life of either one, both are desensitized to later close sexual attraction and bonding. The Westermarck effect has been well documented in anthropological studies, although the genetic prescription and neurobiological mechanics underlying it remain to be studied. What makes the human evidence the more convincing is that all of the nonhuman primates whose sexual behavior has been closely studied also display the Westermarck effect. It therefore appears probable that the trait prevailed in the human ancestral line millions of years before the origin of *Homo sapiens*, our present-day species. The existence of the Westermarck effect runs directly counter to the more widely known Freudian theory of incest avoidance. Freud argued that members of the same family lust for one another, making it necessary for societies to create incest taboos in order to avoid the social damage that would follow if within-family sex were allowed. But the opposite is evidently true. That is, incest taboos arise naturally as products of response mediated by a relatively simple inherited epigenetic rule. The epigenetic rule is the Westermarck effect. The adaptive advantage of the Westermarck effect is, of course, that it reduces inbreeding depression and the production of dead or defective children. That relentless pressure is almost surely how it arose through evolution by natural selection.

The value of the consilience program – or renewal of the Enlightenment agenda if you prefer – is that at long last we appear to have acquired the means either to establish the truth of the fundamental unity of knowledge, or to discard the idea. I think we are going to establish it. The great branches of learning seem destined to meet this way, and if so it will be a historic event that happens only once. But of course we must be careful: surprises, even shocking surprises, may occur. What will be the outcome? Human nature is such that we will find out, and probably in a few years. That prospect is what makes future scholarship in all of the great branches of learning, the natural sciences, social sciences, and humanities, so very exciting and worthwhile. Hence it is also why the World Knowledge Dialogue is a timely initiative.

Introduction

Transforming Knowledge into Wisdom

by Moira Cockell, Jérôme Billotte, Frédéric Darbellay and Francis Waldvogel[1]

Knowledge cannot exist in the absence of dialogue. Whether at the level of socio-cultural interactions, Internet web design or neurological signaling processes, knowledge accrues through the bi-directional exchange of information across networks of receivers and emitters. Without such dialogue there is no transmission, revision or expansion of knowledge because these are inherently dynamic reflexive processes. All knowledge is thus sterile and ultimately doomed to wither unless it is shared, free to mutate and evolve. And like other living systems, the realm of the mind needs cross fertilization to maintain its vigor. That is why this book marks a milestone in constructive exchange between the physical, natural and human science disciplines.

Near the end of 2008, two hundred and fifty academics met together to initiate a dialogue between peers from the different great branches of learning, as part of an experiment in interdisciplinary bridge-building. The common aim: to explore ways of crossing long-held communication barriers between scholars of the natural/technical sciences and the social sciences, arts and humanities disciplines. The WKD Foundation then commissioned a number of the participants to write articles that explore in more depth some of the issues they discussed together. Willing contributors ranged from young doctoral students to venerable senior figures. The resulting articles vary greatly in subject matter, style, length and overall structure. However what emerges from these sixteen original and highly personal essays portraying multiple facets of the knowledge creation process, is a common sense of purpose and a shared framework of new tools and approaches to interdisciplinary dialogue. The authors have not only taken the time to articulate their own perceptions of some complex problems for a broad audience, they have made an effort to consider things from each other's point of view and have responded patiently

1 The authors are past and current members of the World Knowledge Dialogue Foundation's Executive Board. See http://www.wkdialogue.org/

to questions, suggestions and criticisms from each other. They may not wholeheartedly share each other's opinions or choose to take an identical strategic approach to research, but in their willingness to participate in and contribute to the World Knowledge dialogue Foundation's initiatives, each exemplifies the open-minded spirit that is an essential ingredient of creativity.

Do not expect to find here the blueprints for how to foster innovation, achieve universal harmony or share the fruits of knowledge equitably: Such goals are moving targets and as such there can be no finite way to attain them, strive though we should to get closer. To glean the secret of success you must also read between the lines! Each epistemological approach to understanding the universe and our place in it has its limitations and boundaries. However just as individuals have personal blind spots and see more easily the limitations of others than their own, the same is surely true for the different branches of learning. As individuals, we learn to triangulate around our blind spots, stretch our limits and expand our horizons, by assimilating criticism, seeking advice and recognizing the tension between emotional and rational components of our intellect. The same strategy also applies at the larger scale of the different branches of learning, provided these actively communicate with each other and value constructive criticism as a productive rather than destructive force. Interdisciplinarity in this sense becomes an essential component of a balanced education, not merely a fashionable buzzword for the 21st century or a means to access new sources of funding for one's own favorite research.

We live in times where the structure of human society has become so heavily dependent on science and technology, that some fear the consequences of a growing modern trend towards illiteracy in subjects such as mathematics and the physical sciences, while others bemoan a common perception that the pursuit of scientific knowledge now trumps all the other domains of human intelligence. Those who care to can find evidence of both these polarizing effects among certain groups within larger society. Yet the argument, about whether science, or the arts and humanities have contributed most to the history of human development, is a futile one. It is time for us to stop deriding each other and begin to pool our resources. Thanks to scientific, medical and technological breakthroughs, human society's capacity to live longer, migrate further and exploit more of the resources around it has exploded in the space of just a single lifespan. Nonetheless, we encounter daily reminders that science and technology are not an unmitigated force for good. They are merely tools that we can choose to employ. The choice between use or misuse of such tools depends on maintaining a broad sector of educated society that is prepared to engage in ethical, philosophical and political issues and to accept responsibility for deciding what is done with the knowledge we generate.

The advent of global communication among human beings has not elevated us to the realm of god-like creatures; rather, it has bequeathed us a global responsibility for each other. The individual may make the case for his or her pursuit of fundamental knowledge for its own sake, but if the elite group of society that is academia does not embrace the responsibilities that accompany its place of privilege, it may not endure long as a model.

Chapter 1

Common Responsibilities

1.1 Public, Relational and Organizational Trust in Economic Affairs[1]

by Karen S. Cook and Oliver Schilke[2]

One of the most significant economists of our time, Kenneth Arrow, argued over three decades ago that trust has implications for the economy as well as the polity. In the current "economic crisis" his words seem prophetic. For Arrow, trust has not only economic value but also sheer pragmatic value. It simply makes life easier. Like Luhmann, Arrow viewed trust as an important lubricant of a social system: "It is extremely efficient; it saves a lot of trouble to have a fair degree of reliance on other people's word" (Arrow, 1974, p. 23). Note here that the term trust is meant to imply honesty and integrity – that one's word can be counted on.

Arrow argues that trust not only saves on dyadic transaction costs (the concept Oliver Williamson made famous) associated with individual exchanges, it also increases the efficiency of a system. It enables the production of more goods (or more of what a group values if we focus on public goods) at less cost. But, to carry his economic analysis of trust further, it cannot be simply produced on demand and it cannot be bought and sold on the open market. In his words, "it is not even necessarily very easy for it to be achieved" (Arrow, 1974, p. 36). In fact he was not at all sure how it could be produced in societies in which it does not exist. Reflecting a theme that is central to the subsequent work of Frances Fukuyama (1995), almost two decades later, Arrow argued that one of the properties of those societies classified as less developed economically is the lack of mutual trust (or what Fukuyama and others later came to refer to as generalized trust). The lack of mutual trust makes collective undertakings difficult, if not impossible,

---

1 This article has now also appeared in *Corporate Social Responsibility* (2010, Volume 13, Number 2, pp. 98-109).

2 Karen Cook is the Ray Lyman Wilbur Professor of Sociology and current chair of the Department of Sociology, Stanford University. Oliver Schilke is a doctoral student at the Department of Sociology, University of California, Los Angeles and is pursuing his Habilitation at RWTH Aachen University (Germany).

since individuals cannot know if they engage in an action to benefit another that the action will be reciprocated. It is not only the problem of not knowing whom to trust, it is also the problem of having others not know they can trust you.

Arrow's discussion of trust and its economic implications is brief, but brings to the surface some of the fundamental problems with treating markets (and prices) as the main mechanism for coordinating the interests of individuals within a society or at least the limits of price as a governance mechanism. The lack of mutual trust, Arrow points out, represents a distinct loss economically as well as a loss in the smooth running of the political system which requires the success of collective undertakings.

The economic value of trust in Arrow's view thus has mainly to do with its role in the production of public goods. Individuals have to occasionally respond to the demands of society even when such demands conflict with their own individual interests. Certainly, trust has been viewed over the decades as central in the solution of what has come to be known as the prisoner's dilemma (and by extension many social dilemmas). The two prisoners captured and placed in separate rooms by interrogators must trust each other enough not to turn state's evidence on their partner in crime. If they do, both end up with the worst possible outcomes, mutual conviction. If they maintain mutual trust and remain silent, in the classic version of the prisoner's dilemma, they go free, obtaining the best possible outcome. Without trust, each defects independently sending them both to jail for the maximum amount of time. A long tradition of experimental work in social psychology and economics provides evidence of the frequent failure of mutual trust under such circumstances (Cook and Cooper, 2003). As Arrow (1974, p. 26) notes: "the agreement to trust each other can not be bought."

Public Trust

The work on collective trust, generalized trust and, more recently what is called "public trust", that followed Arrow's famous essay continues to investigate the role of trust in the provision of public goods and in public life more generally. Without generalized trust many argue that the collective action problems of our day cannot be solved easily. Ostrom and Walker (2003), in their review of the various solutions to collective action problems, provide evidence of this fact. There are many ways in which collective action problems are solved in different contexts, only one of which relies on the assumption that generalized trust works to solve the incentive problems involved. Without public trust many of the institutions that provide the foundations for the smooth functioning of society are weakened at best if not made wholly ineffective. At worst, lack of transparency and perceived incompetence serve as the seedbed of corruption.

Arrow's brief treatment of trust foreshadowed much later discussions of the role of trust in the economy. Perhaps the most widely read in the 1990's was Frances Fukuyama's (1995) major treatise, *Trust: The Social Virtues and the Creation of Prosperity*, on the economic implications of trust. This book investigates the links between social variables such as trust or reliability and various economic outcomes. He goes so far as to argue that there are major cultural differences in economic success that are based on the levels of what he terms "general social trust" in the societies he considers including, Japan, the United

States, China, France, South Korea, Germany, Great Britain, Italy and Russia. He analyzes some of the factors that support such a claim.

If, as Fukuyama argues, the ability of companies to move from large hierarchies to flexible networks of smaller firms depends on the degree of trust and social capital in the broader society, then understanding how trust emerges and how it varies across cultures is important in the effort to analyze what makes for economic success in different settings. Others are concerned less with economic performance and more with the social and political consequences of different levels of trust in various cultures (e.g. Japan, China, Germany, the emerging capitalist societies in the former Soviet Union, and the United States). In the United States, for example, it is sometimes argued that it is the breakdown in community and the trust it fosters that is associated with increased criminal activity, violence and anomie all of which have consequences for economic enterprises in urban areas. Uslaner and Brown (2005) and others have studied the links between general social trust and various indicators of inequality, which has been increasing in the United States as well as in many other countries and has been argued to be at the core of decreasing generalized trust.

Fukuyama (1995) reasons that it is social trust that generates the conditions under which specific forms of organization emerge that facilitate competitive economic enterprise. Arrow (1974) argues that economic productivity is hampered by monitoring and sanctioning, when these are required for managing relations based on distrust. Both arguments are transactions cost arguments. It is the lack of social trust that Fukuyama identifies as the reason that organizations adopt a more hierarchical form. The more flexible networks of smaller firms that engage in exchange require trust. In Fukuyama's words (1995, p. 25): "A 'virtual' firm can have abundant information coming through network wires about its suppliers and contractors. But if they are all crooks or frauds, dealing with them will remain a costly process involving complex contracts and time-consuming enforcement. Without trust, there will be strong incentive to bring these activities in-house and restore old hierarchies." Traditional hierarchical forms of governance are thus viewed as inimical to modern global economic activity resulting in lower economic performance.

It is precisely the ability to be flexible and to form networks of small companies that can be responsive to change that Fukuyama identifies as central to economic growth and prosperity. Cultures that fit this motif are poised for economic success in the global economy. Ironically, he argues that it is precisely those cultures with strong and large families that have lower social trust and national prosperity. Fukuyama (1995) refers to this claim as the "paradox of family values."

Oliver Williamson views trust as having a much narrower role in the economy, treating trust as largely relevant only in the realm of personal relations and not at all in economic relations that he characterizes as laden with opportunism. His work is at odds with much that has been written on the role of trust in the economy. In some respects Williamson has a more "romantic" view of trust, wanting to limit the term to the situation in which calculativeness is suspended. For Williamson (1993) the concept trust loses it's meaning if it is not restricted to apply exclusively to personal relations. In his view personal and commercial relations are based on completely different logics involving completely distinct forms of underlying calculus. Other social scientists such as Fukuyama make much broader claims concerning the role of trust in society.

Bradach and Eccles (1989), for example, view trust as one type of control system to be distinguished from price and authority, building upon Arrow's early treatment of governance mechanisms. Reliability and flexibility are important aspects of business relations and Bradach and Eccles associate these characteristics with trust relations. Especially under uncertainty trust becomes an important determinant of transactions as exchange partners seek out those who are trustworthy and likely to be reliable in continued exchange.

In related experimental literature, Yamagishi *et al.* (1998), among others, demonstrate that uncertainty leads to commitment among exchange partners as they attempt to avoid opportunism and potential exploitation or defaults. This same phenomenon is called "relational contracting" in an older literature (*cf.* Macauley, 1963). The tendency to form committed relations and to "lock-in" has some associated opportunity costs since committed exchange partners may not explore new relations that might yield better terms. It is this "stickiness" to certain partnerships often created by trust and commitment that may have significant effects on economic outcomes, especially if there are fundamental changes in the economy such as may be created by new technologies and new or rapidly expanding markets for trade and production. There is also the tendency to stick with exchange partners under conditions of high economic uncertainty and risk, which may lead to missed opportunities when economic conditions change.

Sociologists and anthropologists who study the economy have come to conclusions similar to those of Arrow concerning the role of trust in economic endeavors. Trust, when it exists, can reduce various kinds of costs, including, but not limited to, transaction costs and the costs of monitoring and sanctioning. Granovetter (1985), for example, views economic relations as strongly interrelated with social relations. In this view economic transactions are frequently embedded in social structures that are formed by the social ties among actors. A network of social relations thus represents a kind of "market" in which goods are bought and sold or bartered. In addition, they set the terms of exchange sometimes altering the mode of exchange as well as the content of the negotiations. Trust discourages malfeasance and opportunism in part because when transactions are embedded in social relations reputations come into play. Individuals, he argues, have an incentive to be trustworthy to secure the possibility of future transactions. Continuing social relations characterized by trust have the property that they constrain opportunistic behavior because of the value of the association. Hardin's book *Trust and Trustworthiness* (2002, see also Cook, Hardin and Levi, 2005) portrays an encapsulated interest theory of trust, which is also based on this logic.

Relational Trust

Trust can be defined in relational terms as the belief that the trustee will take one's interests to heart. In the encapsulated interest view of trust articulated in Hardin's (2002) book, *Trust and Trustworthiness* and expanded in Cook, Hardin and Levi (2005), A trusts B with respect to x when A believes that her interests are included in B's utility function, so that B values what A desires because B wants to maintain good relations with A. Other's define trust as the belief that the trustee will not take advantage of one's vulnerability.

If I perceive someone as trustworthy I am less likely to monitor her behavior or performance. In this way trust reduces transaction costs.

In a study of physician-patient trust Cook *et al.* (2004a) argue that trust is central to the smooth functioning of professional-client relations in the medical world, as well as in other contexts. But the perception of the extent to which physicians in general fulfill their fiduciary role to patients at large has clearly declined over the past few decades (Imber, 2008). As Imber notes in his lengthy treatment of trust in doctors, the physician in the early decades was not only a healer, but also a "man" of integrity. A good physician was a "good man" who was competent and, importantly, someone who was a person of character and obligation, as immortalized by Hippocrates in what has become the sacred oath that physicians in training take.

One key reason for examining the role of trust in physician-patient relations is to investigate the ways in which trust "saves" not only on transaction costs but, literally, in terms of reductions in the cost of the delivery of care (e.g. a decrease in the duplication of services). Physicians often cited trust in a patient as one factor in their decisions regarding treatment options. In addition, both patients and physicians talked about how trust made it possible for the physician to respond to patient concerns over the phone or by email when they had established a long-term relationship characterized by trust.

Separate from the decline in trust in the "profession" of doctoring, public trust in those who manage and deliver health care has also declined. What we found in our interviews with patients and physicians is that often patients trusted their own personal physicians (especially if they had a longer-term relationship with them) and they sometimes transferred this trust to the group practice level and to the organization in which the practitioners were embedded, but this also depended heavily on the general reputation of the organization. Both past experience and reputation mattered in their assessments of trustworthiness.

In contrast, just one relatively negative experience with a physician in the group led to continued wariness and lack of trust that generalized to the organizational context. While there is no simple way in which interpersonal trust transfers to larger units in which the dyad is embedded, it is one path by which trust can be rebuilt when distrust exists. Both physicians and patients talked at length about the ways in which trust could be established through specific behaviors and about the value of trust for treatment decisions as well as compliance (or adherence to medical regimens). Several patients also spoke eloquently about the ways in which specific physician actions had undermined their trust in all physicians. In such situations it was difficult for these patients to trust any physician. With distrust that has generalized, rebuilding trust at the relational level may be one of the only mechanisms for resolving the underlying trust breach.

It is interesting that individuals often seem to maintain belief in the honesty and trustworthiness of their own physicians even as their confidence in the profession of medicine at large and the leaders in the field has diminished. This is also typically true of evaluations of local politicians in contrast to politicians in general. Organizational mechanisms for ensuring trustworthiness have arisen in part due to this decline in confidence in the profession of medicine (and health care in general). Ironically, the imposition of such mechanisms often undermines trust at the interpersonal level. But despite the rise of organizational mechanisms for ensuring trustworthiness, trust still has an important role to play in physician-patient relations in which one party is more vulnerable (or less powerful) than the other and monitoring can not be fully effective. The role of trust, however, has diminished over time as the stakes for malpractice have risen and economic factors have driven a wedge between patients and their physicians increasing the perception that conflicts of interest exist that undermine trust (Mechanic, 1998). This macro-level fact has had negative implications for public trust in the healthcare system (as current news accounts reveal in the most recent efforts to reform the system in the United States).

There is an important distinction between trust in individuals and trust in the organization in which they are embedded, which is based in part on the nature of the roles individuals play in the organization and the extent to which they fulfill their roles. This is as true of physicians and nurses in the medical context as it is in the business world. Cook, Hardin and Levi (2005) discuss the nature of professional associations and the role they play in maintaining the reputation of the profession. The AMA (American Medical Association) is an example of a proactive professional association that has worked hard to protect the reputation of physicians and their authority over the conditions of practice. At the same time increasing public lack of trust in various professionals, not just doctors but priests, pastors, lawyers and teachers as well, has resulted in increasing organizational controls and monitoring of compliance with professional ethics. When people have little confidence not only in politicians, doctors, and lawyers, but also in business executives who distort stock prices or engage in "creative," but illegal accounting practices attention shifts to the failure of organizational incentive structures and the relevant regulations to restrain opportunism, sanction the untrustworthy and those who are grossly incompetent.

Organizational Trust

Powell (1996) conceives trust as similar to human or moral capital operating distinctly differently than physical forms of capital. The supply of trust, he argues, increases with use rather than decreasing in value. Trust is not depleted in the same way that physical capital is over time when it is used. Powell (1996) identifies a number of types of business networks in which trust plays a role in the organization of economic activity. For example, in research and development networks such as those in Silicon Valley, trust is formed and maintained through professional memberships in relevant associations, a high degree of information flow across the network and by frequent shifting of employees across organizational boundaries. In another example Powell explores the role of trust in business groups such as the Japanese keiretsu and the Korean chaebol. In these business groups trust emerges out of a mixture of common membership in the group, perceived obligation and vigilance. Long term repeat interactions are key to the establishment of trust relations in this context as well as in most circumstances in which trust relations emerge. Repeat interactions provide the opportunity for learning, monitoring, dyadic sanctioning, and increasing mutual dependence which reinforces the basis for trust.

In a study of the garment industry Brian Uzzi (1997) also identifies the nature of the social relations that link economic actors in ways that determine economic performance. Two types of relationships seem to have been common among firms in the manufacturing business, those characterized as close relations and those characterized as arms-length relations. Those connected by "close" relationships were more likely to be trusting and cooperative, even though the same individuals could be self-interested and businesslike in their arms-length relations. In the close relationships the individuals would more often engage in joint problem solving, transfer finer grained information to one another and generally be more trusting. In contrast, the arms-length relationships were more typically economic relations characterized by lack of reciprocity, less continuity and a focus on narrowly economic matters. Trust developed in relations between manufacturers when extra effort was initially offered voluntarily and then reciprocated, in much the same way that Blau, in his influential book, *Exchange and Power in Social Life*, (1964) suggests that trust emerges in social exchange relations. Uzzi notes that this extra effort might involve giving an exchange partner preferred treatment, offering overtime, or placing an order before it was needed to help a partner during a slow time. Trust relations involved less monitoring.

In other industries sociologists have found that trust relations can sometimes impede economic success. For example, Mizruchi and Stearns (2001) examined the role of trust in the closure of bank deals. Under uncertainty they discovered investment bankers turned to customers they had close ties to, involving trust relations. By engaging in deals with their close friends they were less successful in actually completing deals, thus this practice entailed an economic cost to the banking industry. It may have been that the bankers were unwilling to exercise the same degree of authority over their friends in bringing their deals to close. The latitude one extends a friend in this context thus had a negative impact on the profitability of the banks involved.

Trust between partners in an alliance reduces the need for hierarchical controls (Gulati and Singh, 1998). Higher levels of trust among partners in an alliance results in fewer concerns over opportunism or exploitation because the firms have greater confidence in the predictability and reliability of one another. Alliances between firms that view each other as trustworthy lowers coordination costs improving efficiency in part because the firms are more likely to be willing to learn each other's rules and standard operating procedures. Without such trust hierarchical controls and systems of monitoring and sanctioning are more often put into place to implement the alliance and to ensure success, though frequently increasing the overall cost of the enterprise.

Schilke and Cook (2009) study the role of trust in strategic alliances formed between firms primarily for research and development in Germany. They focus on the determinants of trustworthiness. In a recent study data were collected from firms that had been engaged in strategic alliances in various industries including chemicals, electronics, machinery, motor vehicles and information technology. The core firms were asked to identify some of their key partners and data were then collected from the partner firms to provide dyadic data. The eventual sample of firms included dyadic data from 210 managers. Key informants in the R&D units of each firm provided relevant information on these alliances. The final dyadic sample included data on 167 dyads (after attrition at each stage of the study). Below we identify some of the relevant factors that entered the trustworthiness assessments of these firms.

We defined trustworthiness as an attribute of an exchange partner. Conceptually it refers to the extent to which that partner is viewed as unlikely to exploit any vulnerabilities the other partner has. The typical dimensions of trustworthiness identified in the empirical trust literature include ability (or competence), benevolence and integrity. All three dimensions tend to be a part of trustworthiness judgments. Ability refers to the competencies of the trustee that enable it to perform relevant tasks effectively. Benevolence refers to the extent to which the trustor wants the trustee to do well. In personal relationships it refers to generosity and the willingness to help the trustee. Integrity is the extent to which one can rely on the word or promise entailed in the trust relationship. It involves the perception on the part of the trustor that the trustee adheres to a set of principles that are known and that imply integrity. All of these dimensions are attributes of the trustee, while trust is typically a characteristic of a relationship between two entities.

The two primary elements we examined in terms of assessments of the trustworthiness of alliance partners were organizational culture and level of contractual safeguards. Contractual safeguards are stipulations included in the partnership agreement that impose penalties on a party for non-fulfillment of obligations or expected performance or for failure to cooperate as specified in the agreement (Parkhe, 1993). Organizational culture has been conceptualized in many ways, but we used Barney's (1986) definition, which specifies culture as the complex set of beliefs, values, assumptions and symbols that define the way a firm conducts its business. Cameron and Freeman (1991) conceptualized and operationalized four key types of organizational culture (hierarchy, market, clan and adhocracy). Given that clan culture has been argued to be more closely linked to trust (McEvily *et al.*, 2003, p. 92) we measured the extent to which the firms in our sample could be characterized as having a clan culture, defined as emphasizing participation, cohesiveness and teamwork.

Moderating variables included (1) familiarity – the extent to which the exchange partners had had experience with each other, and (2) reputation – the extent to which the public knew about the firm and their perception of its products and prospects in comparison with competing firms in the industry. We reasoned that familiarity and reputation would moderate the effect of organizational culture and the presence of contractual safeguards on assessments of trustworthiness. Both clan culture and contractual safeguards are antecedents of trustworthiness, but their relative impact on the establishment of the trustworthiness of the firm varies by characteristics of the firm. (see Schilke and Cook, 2009 for more details of the study and the findings). In particular our research indicates that when familiarity is high between two partner firms organizational culture is more important as the origin of trustworthiness. In addition, we find that contractual safeguards are more significant understandably when the reputation of the alliance firm is less favorable. Such safeguards may be viewed as increasingly essential in settings, which are more economically volatile. Alliance partners will require it and the public will want such assurance as a precursor to investment decisions (sometimes made with very little information when public trust and confidence in business was much higher).

Under some circumstances research indicates that trust and trustworthiness can improve the workings of organizations and markets. In these instances, however, trust and trustworthiness tend to be complements to structured incentives and to monitoring and enforcement mechanisms (see also Arrow, 1974, p. 24). The backdrop of third party enforcement can give individuals confidence to treat each other as if they are trustworthy at least in those domains where violations of trust will be punished or in which little is at stake. This may enable individuals to learn more about each other, to begin to take risks with each other, and in time to become trustworthy to each other. Examples of third party enforcers abound: Legal institutions that enforce contracts, managers who supervise employee relationships with clients, professional associations that investigate unethical behavior of their members, and hospital boards that inhibit malpractice. They not only boost the probability of reliable behaviors but also create the conditions for cooperation and trust. In an era of decreasing public trust in the world of business such mechanisms may be critical for the reestablishment of confidence and trust in general in economic institutions.

In the world of informal economies, Portes and Sensenbrenner (1993) analyze the role of trust in economic outcomes for immigrants to empirically demonstrate the impact of the embeddedness of economic activities in social relations. In particular, trust plays a big role in the informal economy in which immigrants are able to barter and trade services outside of the formal economy with individuals they deem trustworthy in their personal networks. (Such informal economic transaction systems are rapidly growing as a result of the current economic crisis. Trade and barter of personal items and services now occurs over the Internet and is spreading globally – e.g. Freecycle.org.) Immigrants also use their social networks as a kind of social capital when they enter a new country to provide access to critical resources such as educational and training opportunities, entry jobs, and the provision of food and shelter until they can become established on their own terms. The social networks provide the social capital the immigrants need to get established in a new land. Some of these network ties represent trust relations – others do not; thus, it is important to distinguish between trust and social capital. There are also downsides to the use of social networks for immigrants. Closed networks may result which lock the employees into low wage jobs with little time to develop the human capital that would be needed to move up and out of the protective environment of their enclave.

In an interesting study of the U.S. economy between 1840 and 1920 Lynne Zucker (1986) identified three basic modes of trust production in society. First, there is process-based trust that is tied to a history of past or expected exchange (e.g. gift exchange). Reputations work to support trust based exchange because past exchange behavior provides accurate information that can easily be disseminated in a network of communal relations. Process-based trust has high information requirements and works best in small societies or organizations. The second type of trust she identifies is characteristic-based trust in which trust is tied to a particular person depending on characteristics such as family background or ethnicity. The third type of trust is institutional-based trust, which ties trustworthiness to formal societal structures that function to support cooperation. Such structures include third party intermediaries and professional associations or other forms of certification that remove risk. Government regulation and legislation also provide the institutional background for cooperation lowering the risk of default or opportunism. High rates of immigration, internal migration and the instability of business enterprises from the mid 1800's to the early 1900's, Zucker argues, disrupted process based trust relations.

The move to institutional bases for securing trustworthiness was historically inevitable. Studies by Greif *et al.* (1995) and other economic historians of the emergence of various institutional devices for securing cooperation in long-distance trade in much earlier periods support this claim. Such devices seem to be the focus of political and public attention in the current period of economic crisis and the reemergence of instability in businesses including large banking and investment institutions formerly viewed as highly stable and thus worthy of long-term investment. Restoring confidence in these institutions will require a lot of work by politicians as well as those in the business world. It is not at all clear that past mechanisms for ensuring trustworthiness will garner the public trust after such cataclysmic losses and the attendant unemployment that has affected all sectors of the economy.

A number of economists and sociologists seem to agree that trust does play a role in the economy, but precisely how and in what ways is still under investigation. In Susan Shapiro's book, Wayward Capitalists, (1984) trust is viewed as the foundation of capitalism. Building on the work of Macaulay (1963) and others she argues that financial transactions could not easily occur without trust because most contracts are incomplete. This theme is reflected in much of the work on contracts and is the reason for many alliances and cooperative agreements that build trust between the relevant parties. In significant ways trust can be said to provide the social foundations for economic relations of exchange and production. Monitoring is often ineffective. Sanctioning can be costly. Transactions costs can be high. To the extent that actors are trustworthy with respect to their commitments such costs can be reduced within organizations and in the economy more broadly. But without the institutional backing of contract law and other forms of legal protection few societies rely strictly on the vagaries of personal relations. This seems only to happen when institutional backing is weak or non-existent and when interpersonal relations are the primary locus of exchange (often in developing countries suffering under political instability and corruption). In economies under transition from one major form of economic organization to another, as in the transitions that have occurred in post-communist societies, reliance on personal networks and trust relations can serve an important step in the evolution to systems of trade that require interactions with strangers in the context of market economies. This transition, however, can be highly problematic (Radaev, 2004; Cook, Rice and Gerbasi 2004b; etc.) and can be fraught with risk as in the case of Russia.

Conclusions

Restoring public trust in economic enterprises and in other arenas in society in which trust has declined (e.g. the world of professions including doctors, lawyers, priests and politicians) will not be easy. We have tried to articulate in a few domains how personal relations of trust, organizational level trust, and general social trust are linked. But there is more work to be done on this topic, theoretically, methodologically and empirically. Cook, Hardin and Levi (2005, p. 196) argue that: "Societies are essentially evolving away from trust relationships toward externally regulated behavior." This is in part due to the change in the ways in which we relate to one another. We have evolved over long periods of time away from thick relations of trust and normative control in small communities to larger networks of thin relations of trust and cooperation with many people spread out in geographic space. This has altered the fundamental ways in which business is accomplished and has implications for the potential for trust relations, assessments of trustworthiness, and modes of cooperation. It also has implications for the capacity of the public to know enough to "trust" any institution in the traditional sense. The best we might accomplish is to regain confidence that these institutions are being given proper oversight and that there is legal recourse for those who are the victims of exploitation. A major difficulty is that the scale of business operations has become so complex and interrelated in many sectors that providing such oversight is increasingly complicated, if not impossible. In such a world, public trust in institutions may continue to be fairly low, even though at the personal level (my banker or doctor) or even at the organizational level (my local bank or group medical practice) there may be room for increased trust, given proper organizational incentives. Further research on these complex issues in an increasingly interdependent and global world of economic activity is needed.

References

ARROW, K. J. (1974) *The Limits of Organization*, New York, W.W. Norton and Company.

BARNEY, J. B. (1986) "Organizational Culture: Can it be a Source of Sustained Competitive Advantage?", *Academy of Management Review*, 11 (3), p. 656-665.

CAMERON, K. S. and FREEMAN, S. J. (1991) "Cultural Congruence, Strength and Type: Relationships to Effectiveness," in WOODMAN, R. and PASSMORE, W. A. (Eds), *Research in Organizational Change and Development*, Vol. 5, p. 23-58, Greenwich, CT., JAI Press.

COOK, K. S. and COOPER, R. M. (2003) "Experimental Studies of Cooperation, Trust and Social Exchange", p. 209-244, in *Trust and Reciprocity: Interdisciplinary Lessons for Experimental Research*, edited by OSTROM, E. and WALKER, J., New York, Russell Sage Foundation.

COOK, K. S., KRAMER, R., THOM, D., BAILEY, S., STEPANIKOVA, I. and COOPER, R. (2004a) "Physician-Patient Trust Relations in an Era of Managed Care," in *Trust and Distrust in Organizations*, edited by R. KRAMER and COOK, K. S., New York, Russell Sage Foundation.

COOK, K. S., RICE, E. R. W. and GERBASI, A. (2004b) "The Emergence of Trust Networks under Uncertainty: The Case of Transitional Economies – Insights from Social Psychological Research," in *Problems of Post Socialist Transition: Creating Social Trust*, edited by ROSE-ACKERMAN, S., ROTHSTEIN, B. and KORNAI, J., New York, NY, Palgrave Macmillan.

COOK, K. S., HARDIN, R. and LEVI, M. (2005) *Cooperation without Trust?*, New York, Russell Sage Foundation.

FUKUYAMA, F. (2005) *Trust: The Social Virtues and the Creation of Prosperity*, Free Press, New York

GAMBETTA, D. and HAMILL, H. (2006) *Streetwise: How Taxi Drivers Establish Customers' Trustworthiness*, New York, Russell Sage Foundation.

GANESAN, S. (1994) "Determinants of Long-Term Orientation in Buyer-Seller Relationships," *Journal of Marketing*, 58 (2), p. 1-19.

GRANOVETTER, M. (1985) "Economic Institutions as Social Constructions: A Framework for Analysis," *American Journal of Sociology*, 91, p. 481-510.

GREIF, A., MILGROM, P. and WEINGAST, B. R. (1995) "Coordination, Commitment and Enforcement: The Case of the Merchant Guild," p. 27-56 in *Explaining Social Institutions*, edited by J. KNIGHT and H. SENED, Ann Arbor, University of Michigan Press.

GULATI, R. and SINGH, H. (1998) "The Architecture of Cooperation: Managing Coordination Costs and Appropriation Concerns in Strategic Alliances," *Administrative Science Quarterly*, 43, p. 781-814.

HARDIN, R. (2002) *Trust and Trustworthiness*, New York, Russell Sage Foundation.

IMBER, J. (2008) *Trusting Doctors*, New York, Palgrave.

MACAULAY, S. (1963) "Non-Contractual Relations in Business: A Preliminary Study," *American Sociological Review* 28, p. 55-67.

MECHANIC, D. (1998) "The Functions and Limitations of Trust in the Provision of Medical Care," *Journal of Health Politics, Policy and Law* 23 (4), p. 661-686.

PARKHE, A. (1993) "Strategic Alliance Structuring: A Game Theoretic and Transaction Cost Examination of Interfirm Cooperation," *Academy of Management Journal*, 36 (4), p. 794-829.

PORTES, A. and SENSENBRENNER (1993) "Embeddedness and Immigration: Notes on the Social Determinants of Economic Action," *American Journal of Sociology* 98, p. 1320-1350.

Powell, W. W. (1996) "Trust-Based Forms of Governance," p. 51-67 in *Trust in Organizations: Frontiers of Theory and Research*, edited by Kramer, R. and Tyler, T. R. Thousand Oaks, CA, Sage Publications, Inc.

Radaev, V. (2004) "How Trust is Established in Economic Relationships: When Institutions and Individuals are not Trustworthy," in *Problems of Post Socialist Transition: Creating Social Trust*, edited by Rose-Ackerman, S., Rothstein, B., and Kornai, J., New York, Palgrave Macmillan.

Schilke, O. and Cook, K. S. (2009) "How do Firms Determine the Trustworthiness of an Alliance Partner? The Moderating Effects of Familiarity and Reputation," Unpublished Working Paper.

Uslaner, E. M. and Brown, M. (2005) "Inequality, Trust, and Civic Engagement," *American Politics Research*, 31, p. 868-894.

Uzzi, B. (1997) "Social Structure and Competition in Interfirm Networks: The Paradox of Embeddedness," *Administrative Science Quarterly* 42, p. 35-67.

Williamson, O. E. (1993) "Calculativeness, Trust, and Economic Organization," *The Journal of Law & Economics* XXXVI no. 1 2, p. 453.

Yamagishi, T., Cook, K. S. and Watabe, M. (1998) "Uncertainty, Trust and Commitment Formation in the United States and Japan," *American Journal of Sociology* 104, p. 165-194.

Zucker, L. G. (1986) "Production of Trust: Institutional Sources of Economic Structure, 1840-1920," p. 53-112, in *Research in Organizational Behavior*, edited by Staw, B. M. and Greenwich, L. L., CT, JAI Press Inc.

1.2 Learning from the Past: How to bring Ethics and Economics in line with the real Nature of the Human Being

by Philipp Aerni[1]

Most policy experts failed to anticipate the food and the financial crises in 2008. Meanwhile, many comments on the global economic downturn in 2009 made by economists and ethicists did not go much beyond denouncing the usual suspects in business and politics. Yet, there is increasing evidence that policies derived from their theories that still rely on an ideal rather than the real nature of the human being may have caused the crisis in the first place. This article argues that the economic, social and environmental challenges we face in the 21st century ask for a more interdisciplinary approach in academic research and training, especially in the fields of ethics and economics. These disciplines need to adjust themselves to new insights in the cognitive sciences and anthropology. Moreover they need to become more engaged in field research that begins and ends in the domain of concrete observations, combining the basic elements of inductive and deductive research. This type of holistic research is not new but has its roots in Renaissance Italy. At that time, it was largely a public-private partnership project that enhanced moral imagination, fostered human empowerment and produced useful knowledge that effectively addressed the needs of the people. At that time Europe had to catch up with more advanced economies in the Levant. In the 21st century the emerging economies are in this position and they seem to embrace or at least appreciate more the virtues of the Renaissance entrepreneur. The way they pursue their self-interest in the global knowledge economy also forces them to extend their moral imagination to other cultures. It is in the personal interest of the entrepreneur now and then to focus on what he or she has in common with other people rather than what makes him or her unique. Extending moral imagination may well be the result of natural selection in the process of cultural evolution and thus rooted in science, economic development and religion alike.

1 Philipp Aerni is a Senior Research Fellow at the Institute for Environmental Decisions at ETH Zurich and the World Trade Institute at the University of Bern, Switzerland.

The crisis and what it reveals about the academic system

The food crisis and the subsequent financial crisis in 2008 reveal an intellectual crisis in academia that has grown steadily since the end of the Cold War. These crises are rooted in the fact that institutions that were originally designed to serve the people have converted into large self-contained systems of thought that have developed their own internal logic and thus become increasingly concerned with their own survival and expansion. Simone Weil called this phenomenon 'the great beast' that is intrinsic to every form of evolving social organization (Weil, 1986). However, once these self-contained systems become part of the problem rather than part of the solution to contemporary problems, new opportunities for intellectual renewal emerge. That said, policy strategies pursued before and after the crisis year of 2008 still reflect the ideas that have their roots in the 1970s, ignoring the reality of the global knowledge economy and new insights from recent empirical interdisciplinary research.

It is therefore not surprising that the bureaucracies in finance, government, academia and civil society are mostly in denial about their particular responsibility for the crisis and are engaged in ex-post rationalizations of their policies and recommendations prior to the global crisis.

The case of the food crisis

When world market prices for major food and feed stocks rose to unseen levels in spring 2008, it caused violent protests among the poor in many large cities of the developing world. Their tight household budgets were no longer sufficient to cover the basic expenses for food. This development took policy makers across the globe by surprise because it was widely assumed that the trend of declining food prices over the past decades would continue due to technological progress and the fact that in economic terms food is classified as an "inferior good with a low income elasticity" (the share of food expenses in household budgets normally decreases with increasing income).

When it came to staving off the food crisis, governments largely became concerned with feeding their own people by imposing export restrictions and approving the expansion of cultivated land. The imposition of export restrictions tended to further aggravate the situation for food importing countries. At the same time the expansion of cultivated land manifested itself primarily in the form of large-scale acquisition by governments and large corporations in industrialized and large emerging economies, of land in poor developing countries with a large potential for agricultural development. Even though there might be some positive spillovers for local people, mere extension of agricultural land is hardly an original response and if it requires the clearing of large forests it is also problematic from an environmental perspective (*The Economist*, 2009a). The same lack of originality can be observed with large NGOs concerned with issues related to environment and development. As it is the case with many traditional welfare economists they seem unable to spot any potential of new technologies in addressing the basic problems of food crisis (IAASTD, 2008).

The situation of established welfare economists in general and agricultural economists in particular is that many of them were trained back in the 1970s on highly reductionist and comparative-static equilibrium models. These tended to argue that technical progress largely benefits the consumers by means of lower food prices but not the farmer who would be stuck in the so-called agricultural treadmill; forced to adopt new technologies in an environment of cost-cutting perfect competition but thereby contributing to productivity increases that lower food prices due to the increase in supply and the low uncome elasticity of food products (Cochrane, 1979). Furthermore, they followed old growth theory in assuming that knowledge is a non-rival and non-excludable public good that is produced exclusively in the public sector (Solow, 1957). Reality however clearly shows that the private sector invests almost as much if not more in agricultural R&D than the public sector and by means of patenting knowledge they make it a partially excludable good. Agricultural R&D knowledge is therefore not a pure public good, yet the eventual commercialization of this knowledge in the form of new goods and services produces large welfare effects that neoclassical welfare economics largely ignores because they do not fit into the comparative-static equilibrium models (Romer, 1994). NGO activists committed to sustainable development share the economist's view to the extent that they believe that technical progress in agriculture produces negative externalities and therefore market failure that needs to be addressed by the public sector. Mostly trained in the social sciences or humanities in the 1970s, these NGO experts further share with the economist a belief in an ideal rational human being. But instead of the economists' assumption that the rational human being pursues his self-interest without any other-regarding preferences (the sociopath called homo economicus), they believe in an equally absurd human being who is guided by universal and reason-based ethical principles that enable him to become a perfectly responsible citizen in a globalized world. Within this simplified view, the task of NGOs would be to educate and support these citizens by alerting them about the current state of the world and how capitalist civilization and technological progress are undermining the very resource base that they depend on in order to survive. Yet, the problem with this view is that NGO activists themselves are not really doing serious research on the state of the world but selectively pass on academic claims that match the preferences of their donors and confirm their particular world view (Herring, 2009).

The food crisis put many NGOs in an awkward position because according to their world view, there is enough food around if only it would be better distributed. Therefore they preferred to interpret the food crisis as a consequence of the waste of food in the industrialized world, speculation with food prices and the promotion of biofuels. Moreover, they insisted that technological innovation, particularly in the area of agricultural biotechnology, cannot contribute to the solution of the problem since the resulting technologies would just serve large corporations while increasing the risks for resource-poor farmers (Rosset, 2009). Even though a large amount of empirical research contradicts this claim (The World Bank, 2007), it was very convenient that another World Bank-sponsored report on the state of agriculture (IAASTD, 2008) appeared in spring 2008 when the food crisis became acute. The latter report argued that good agricultural practices (meaning organic farming and other labor-intensive and low-productivity practices of integrated pest and soil conservation management) would be the way forward rather than investing in agricultural modernization in the most vulnerable and underdeveloped countries. Its authors were basically using the same arguments that were mustered against the Green Revolution in the 1970s and 80s. Moreover, by endorsing the argument that it is a distribution and not a production problem, they indirectly imply that we could solve the food crisis by donating the agricultural surpluses produced in subsidized rich agricultural systems to food-importing developing countries and thus achieve a better distribution of food. As is well known from actual experience, such a policy would worsen the situation in these developing countries because it would further destroy the economic base of local farmers who cannot compete with free food from abroad. By indirectly endorsing this sort of food dumping and denying the importance of new technologies in addressing not just agricultural but also environmental problems, the reports authors reveal a lack of moral sensitivity for the reality on the ground and a preference for the cherished world views and prejudices of their own constituencies at home, mainly consisting of urban dwellers who lack the ability to understand the different circumstances and priorities that the rural poor face in their struggle to improve their livelihoods. Taking into account recent insights in neuroscience and evolutionary psychology (Striedter, 2004), one might argue that our brains have evolved to empathize only with those who share similar values. Yet, in a globalized world, survival strategies require people more and more to empathize with those who do not share the same culture and belief systems.

The reality on the ground, particularly in Africa, has worsened considerably over the past decade. Sub-Saharan Africa has become more dependent on food aid than ever before (Masters, 2008). One reason for this misery might be the continent's strong ties to its former colonial masters in Europe and the high economic dependence on aid and European market access for agricultural goods. This has made African governments more prone to adopt policy prescriptions from Europe that were based on outdated academic theories and lacked a proper understanding of the local context. The Europeans were also successful in promoting these policies throughout international organizations as a result of their importance as donors in development assistance (Aerni, 2008).

The food crisis of 2008 should nevertheless undermine such neocolonial development strategies, especially considering the fact that, for more than a decade, South-South collaboration within the developing world is growing much faster than North-South collaboration. South-South collaboration is less concerned with standards and regulations than it is with business facilitation. Its primary purpose is to jointly address economic and social problems, by jump-starting joint R&D programs designed to promote home-grown technological development with lots of spillover effects for the local economies (*The Economist*, 2009b).

Maybe South-South collaboration is not addressing the problems that are high on the priority list of Europeans and Americans, but they seem to be more effective in meeting the goals they intend to achieve because the South-South networks are far more pragmatic in their approach. Furthermore they do not have to care about pleasing the people back home as it is the case with many donor agencies, NGOs and supermarket chains based in Western countries, and they are generally less ideological. They simply cannot afford to waste money by organizing countless local workshops in the name of 'capacity building', (a term that frequently serves to impose foreign views on local people) and then call it 'empowerment' or 'help for self help' (Aerni, 2009). With the global economic downturn and the urgency to effectively address the global environmental challenges, the justification for such expenses may also have to be revisited by European donors. There is a need to focus more on the priorities of the poor and to take more seriously what they consider to be major problems and how these ought to be addressed. This is of course a challenge, not least because the aid-dispensers often feel that they must know better due to their 'superior' education received in their home country. They frequently attribute their referential treatment in the host country in Africa to this 'superior' education rather than to the fact that their visit might bring money; yet not surprisingly, aid-recipients are often willing to agree with everything proposed as long as aid money is dispensed. Once Western donors realize that the best ideas in development practically never come from idealists in Europe or the US but rather from local practitioners in developing countries themselves, they might change their minds. Simply put, the major difference is that while Western idealists see the problems and want to solve them by changing the system, local practitioners see the opportunities within the existing system and want to realize them. The global knowledge economy indeed offers a lot of opportunities for local people to combine their practical local knowledge with the use of new technologies that are sufficiently affordable and flexible to contribute to tailor-made solutions (von Hibbel, 2006). In this context, it is not important whether the development is produced in the South or the North, or the private sector or the public sector, as long as it is useful and applicable. These are lessons that can be learned from the pragmatic approach in South-South research collaborations where stakeholders have learned from experience that the only non-scarce resource, the only resource that is not depleted with increasing use, is knowledge. Therefore major investment needs to go into the production and use of knowledge that is of practical relevance. This means investing in human capital and the mobilization of science and technology for cost-effective technological innovations that foster economic development and a more sustainable use of natural resources. In this endeavor, we should abandon our cherished antagonisms in which one has to take sides between small-scale and large-scale agriculture, traditional knowledge and new technologies, public sector activities and private sector activities, etc. In the face of the urgency to address the complex problems related to hunger, malnutrition and environmental degradation it is pointless to cling to such a constructed bipolar world at the expense of tapping all the possible resources and people available to address local and global problems.

It would be especially important to get the private sector more involved in the joint undertaking to promote agricultural sustainability (rather than alienating it as the IAASTD report did). Companies that invest in new knowledge in order to generate new products and solutions not just for the affluent but also for the poor should be awarded accordingly. The private sector should be lured with special incentives to participate in local public-private partnerships (PPPs) for sustainable development. Companies are quite predictable in their calculations. If the potential benefits of partnership exceed the potential costs they will participate. Neoclassical welfare economics is not dealing with potential welfare effects that could be gained through PPPs because its social planning approach implies that only public sector activities can produce welfare effects (Romer, 1994). It is therefore not surprising that economists mostly recommended top-down policies that hardly ever addressed the real needs of the poor through their integration into the global knowledge economy. The failed integration especially of the rural poor in the process of globalization has fatal consequences because it increases their dependence on outside support and makes them more vulnerable to external shocks.

The financial crisis and the subsequent economic downturn

It may be a coincidence that the food crisis and the financial crisis occurred in the same year, but both crises are rooted in same flawed comparative-static equilibrium models that ignore the evolutionary character of every economic system (Beinhoecker, 2006). Moreover, modern economies tends to ignore that it is unconscious or bodily thinking rather than the rational conscious mind that is responsible for the most part of human decision-making (Grün, 2009). These "animal" spirits account for the often messy decision-making process of economic agents (Akerlof and Shiller, 2009). They were still taken into account by the great economists in the early 20th century but somehow disappeared from the mainstream textbooks in economic theory.

For many decades, general equilibrium (GE) models are applied by agricultural economists working in institutions such as the World Bank and macroeconomists working in central banks. These economists mostly worried about the prices of goods and services but neglected the prices of assets, trusting the financial markets to properly assess them. In such GE models bankruptcy cannot occur because financial intermediaries like banks do not appear in the model. Convenience led economists to abide by these models, believing that what does not appear in the model must be irrelevant (*The Economist*, 2009c).

The academic discipline of financial economics has seemed to espouse almost a religious faith in models that assume purely rational economic agents. Most of these models are based on the efficient-market hypothesis (EMH) developed by Eugene Fama in the 1970s at the University of Chicago. The EMH argued that the price of a financial asset always reflects all available information that is relevant to its value. Wall Street then concluded that markets would automatically price financial assets accurately. Deviations from equilibrium values would be quickly detected by the rational investor making money by exploiting the deviation by means of the various financial instruments available. The theory was that such rational investors would also quickly detect a bubble and prick it before it grew out of scale. In the current climate it is no longer necessary to rebut these assumptions in view of their thorough falsification through the financial crisis. But we should not lose sight of the fact that most public policy instruments today are still based on the idealist view of a purely rational and cognitive human being.

In short, the food as well as the financial crisis revealed that the economic models that rely on purely rational economic agents may do more harm than good in public policy. The human being is guided by a dynamic belief system that largely reflects life experience as well as internalized knowledge acquired from different sources. The thin view of the human being as a human oeconomicus that is driven exclusively by material self-interest us well as exogenously defined and stable individual preferences contradicts all insights from experimental psychology and neuroscience. Moreover, the human is generally not a passive observer but one actively engaged in pattern recognition and storytelling to make sense out of this world and find opportunities to make a living (Beinhoecker, 2006).

How ethics and economics have lost touch with real human nature

The deductive approach preferred by traditional economists does not permit such considerations. It starts from a set of premises (the perfectly rational economic agent is one of these premises) and then tests them empirically, mostly by feeding the models with secondary data from large databases. More often than not, the predictions of these models turn out to be inaccurate or simply wrong but explanations are always at hand to explain away inconsistencies (*The Economist*, 2009c).

However, behavioral economists and political economists who generally are more open to interdisciplinary research methods have considered the research results obtained from other social sciences disciplines and the neurosciences. They have introduced a more inductive approach by designing controlled laboratory experiments.

Their findings largely confirm what has already been inferred from insights gained in the cognitive sciences: people may pursue their self-interest but they also have other-regarding preferences that often keep them from maximizing their immediate material benefits. Experiments on cohorts of student volunteers from across the globe seem to confirm a universal preference for fairness and reciprocity in different economic games (Camerer, 2006). Yet, rather than using evolutionary economics and attempting to explain these other-regarding preferences as innate human instincts that evolved during the period of hunter-gatherer communities to facilitate effective cooperation in small-groups, behavioral economists tend to attribute it to another form of rationality. This rationality is supposedly related to universal ethical principles and notions of justice and fairness rather than economic principles. This idealist assumption resembles Kant's view that the real value of a moral act can only be seen in the categorical absence of personal desires and interests. It is also strongly attached to the theory of the stages of moral development that was developed by Lawrence Kohlberg in the 1970s and defines the highest stage of moral reasoning in accordance with the purely cognitive and reason-based view of deontological ethics.

Yet, behind the allegedly objective, unbiased and disinterested reasoning about justice and fairness, there almost always lurks something that has nothing to do with ethics (Haidt, 2007). For example; if we have the power to hurt someone who humiliated us earlier when we were powerless, we will probably do it, but call it 'justice' rather than 'vengeance'. If we see a very big and expensive car parked in front of a village restaurant, we might complain that material inequality is unfair and should be addressed through redistribution. But our underlying feeling is probably closer to personal envy than a genuine desire for fairness. We are however afraid of admitting such motives since our consciousness has evolved to deny the influence of base unconscious feelings on our conscious thinking. Moral fear rather than reason is therefore revealed through our moral systems. This fear is however masked by the taboos and hierarchies of society (Grün, 2009).

Paying attention to the laws of cultural evolution

There are natural forces in the human being that are nevertheless conducive to amity and tolerance as opposed to the forces conducive to belligerence and intolerance. Such forces may indeed reflect a moral truth that is consistent with human nature. The notion of moral truth is however more a product of cultural evolution rather than natural evolution (Wright, 2009). For example, the ability to argue persuasively that your rival had no valid grounds for grievance would have been favored by natural selection in the age of hunters and gatherers when outsiders represented mainly rivals that competed for the same resources. Yet, once the human being started to engage in trade and technological innovation and thus jump-started cultural evolution, zero-sum games transformed into non-zero sum games in which it suddenly made sense to engage with former rivals and exchange goods and services for the benefit of both parties. This transformation also required a change in the evaluation of the moral status of the former rival and the ability to look at things from his point of view.

The first great expansion of non-zero sum games happened around 1000 BC when the Phoenician cities in the Levant started to establish a network of trading points along the shores of the Mediterranean Sea. Phoenician merchants formed cosmopolitan trading companies where Jews, Greeks and other ethnicities shared one ship for a period of about four months to sell their goods in other parts of the world and bring foreign goods back to their home country. The main objective of the Phoenician merchant may have been to convert his service into a profit by exploiting the price differences between center and periphery in the new trading space. However, the moral by-product of this undertaking was better access to knowledge, goods and services for people who previously suffered from physical shortages, a lack of opportunities and few possibilities to change their situation. Apart from that, people from one culture learned about the habits, institutions and languages of people from another culture, and the joint pursuit of a business forced them to respect the other point of view and seek compromise rather than confrontation (Sommer, 2005; Aerni, 2007).

The resulting extension of moral imagination from friends to rivals is therefore designed to steer participants through the successful playing of games – to realize the gains of non-zero-sum games. It then makes suddenly sense to see people not primarily as potential enemies but as people one can do business with (Wright, 2009).

Even negative-sum prospects that may result in the form of side-effects from the proliferation of non-zero games such as increasing social inequality, depletion of the natural resource base, climate change, environmental degradation, etc. give entrepreneurial innovators an incentive to make better use of available knowledge to come up with new technologies, services and policy instruments that help address these challenges through technological and intellectual renewal. Global problems can therefore not be solved by simply physically destroying the existing social order and replacing it by a new one that, over time, will look increasingly similar to the old one. Instead they must be addressed by reforming it. Past evidence indicates that successful reforms always reflect an adjustment process to changing circumstances. They are achieved not just by means of regulating unsustainable change but also by facilitating sustainable change through the mobilization of science, technology and innovation for development (Aerni, 2009).

Linking ethics to religion again

One could argue that even religion plays its part in this adjustment process since it turns out that the religions that have survived best in cultural evolution were those that linked individual salvation to social salvation. This may have its earliest traceable root in the religious beliefs of ancient Egypt: at the court of the god Osiris in afterlife, the deceased was supposedly checked by placing his heart on a scale and balancing it against a feather that represented Maat, the goddess of truth. The crucial test was to measure the truthfulness of the claim that the individual respected the lives and properties of others, including servants and the poor. Maat itself was a kind of metaphysical substance composed of truth, order and cosmic harmony. Osiris acted as the guardian or *lord of maat*, while Seth, the god who killed him and over whom Osiris triumphed through resurrection, acted as the god of chaos (Redford, 2002). Thus the human lives before the court of Osiris had to show how they helped to prevent social dissolution (the forces of Seth) by showing respect for the lives and properties of others.

The dual struggle represented by the forces of Osiris and Seth were subsequently incorporated into the Abrahamic religions. It was rooted in the real challenge of life that in order to avoid individual chaos one had to avoid social chaos (Wright, 2009). In this sense, the message inherent in all the great religions is to find personal salvation by recognizing the human nature that we all share and obey to the necessity of life (Weil, 1986). This necessity comprises the obligation to fulfill one's own potential in order to become productive in life and contribute to the flourishing of the society on which our well-being depends (Spinoza, 1677).

The virtues of Renaissance Florence

One moment in the canon of history when the search for personal salvation and fulfillment coincided most with social salvation was during Renaissance in Italy in general and Florence in particular in the 14th and 15th century. Even though this was an age of political, economic and cultural turbulence, it opened a unique window for the underprivileged to improve their livelihoods through the development of unique skills and hard work. The reason why both human creativity and social welfare thrived most in Florence was probably due to the fact that the privileged classes found it more difficult to maintain their superior social position by means of mere oppression of the lower classes. As Goldthwaite (2009) reveals in his book on the Economy of Renaissance Florence, this was largely related to the political and economic crises of the 1340s.

Before and after 1340, the ruling classes, consisting largely of established bankers and industrialists in the textile industry, were confronted with serious economic problems such as major bankruptcies, a sharp devaluation of the florin due to a steep fall in the ratio of gold to silver and a rebellion of the wool workers (the Ciompi revolt). At the same time they also faced political troubles such as the war against Lucca from 1336-8, tyranny of the Duke of Athens (1342-43), and the victory of the popular government over the oligarchy (1343). In addition to these social turmoils, a shortage of consumer demand and labor supply in the economy due to one of the most severe famines in 1347 and then the Black Death, made things even worse.

Crisis and renewal

This period essentially weakened the power of the ruling guilds in their capacity to exclude new innovative players from entering their business. It also led to new economic and political institutions that favored empowerment through entrepreneurship across all social strata.

These institutions were not designed by a well-meaning social planner, as neoclassical welfare economics assumes has a responsibility to do, but were a by-product of the struggle of the underprivileged for more political and economic participation in the city. Yet, the parties involved did not see the struggle as a zero-sum game, where one party wins and the other loses, but favored formally established political and economic platforms of negotiations where compromises could be reached in mutual respect. The poor did not primarily insist on the redistribution of wealth and power but rather the creation of economic opportunities that would allow them to improve their economic situation gradually, to better invest in the health and education of their offspring and to ensure that they would have more possibilities to fulfill their potential. The pressure resulted in the establishment of countless elementary and abacus schools where the youth learned how to read, write and calculate. Moreover, everyone learned the basic principles of accounting in abacus schools.

The poor are more than blessed objects of charity

After the uprising of the wool workers, known as the Ciompi revolt in 1347, the poor were regarded with more suspicion by the ruling elite. This partly explains why charitable institutions such as the Ospedales and confraternities that were largely funded by the guilds to give charitable support to the less privileged members, went into decline in the second half of the 14th century. It is therefore all the more surprising that the period that followed the upheaval is characterized by relative social and political stability compared to other regions in Europe. Obviously treating the poor with some suspicion rather than as blessed objects of charity also implies to take them more seriously as players who collaborate and compete with the rich in many public institutions. If the privileged treated the underprivileged with suspicion they were probably also more likely to keep the level of discontent among them as low as possible by integrating them better into the economy. This integration was possible because even the poor were now literate and able to do basic accounting. It became more attractive for companies and artisan workshops to hire poor but educated young people. The master-apprentice relationship gave them additional valuable on-the-job training. Such training involved not just basic business practices and techniques in the particular field but also joint trips to other cities nearby or even work in a branch in a foreign country. This made the trainees more familiar with other cultures and languages, which enabled them to learn and understand different points of view. Even though they could not count on formal employment contracts they were still partially protected from abuse by the guilds and the formal political institutions in Florence. Moreover, the fact that they had to work for several different masters in several different fields gave them more versatile professional skills and a higher sense of autonomy. The most important aspect of this education was, however, that they were always in personal contact with the rich of Florence. The underprivileged did not despise entrepreneurial spirit and were very much concerned in their economic and political activities to foster a spirit of merit-based pride and self-confidence among all the citizens of Florence.

The tools of empowerment

The tools of empowerment that created social mobility and an unseen productivity and creativity among its citizens was therefore a social network that was supportive of the young and skilled people who wanted to set up their own business. Moreover, there were economic institutions in place that allowed these skilled people to get access to credit, investment and well-trained employees.

One of the important economic institutions that made entrepreneurship attractive was the concept of partnership. Partnerships (compagni) were a fully evolved institution at the end of 14th century. Men joining together as compagni (merchant bankers, artist-investors, etc) drew up articles of association according to which they agreed to contribute capital toward a business venture that was to last usually from three to five years. The articles of association would specify each partner's contribution to the capital and how his share of profits was to be determined, etc. The partnership contract was a private instrument, not a notarized document.

Public-private partnerships to facilitate social and economic development

These small companies had no need for fixed capital, neither property nor equipment. The head office of the firm was presumably located in the residence of the principal investor, and branch offices and warehouses were rented. Partners could commit further capital on specific terms (e.g. time deposits from outsiders).

The fixed capital necessary to introduce a new product or technique in the market required major investors, however. The Medici regarded the holding company as an appropriate tool for that purpose in the 15th century. By using one partnership to invest in another, the original investors clearly increased the capital they controlled, but they hardly exploited this possibility. As the leading international merchants, the Medicis were also investing in foreign companies that seemed to be highly innovative in the production of high-value goods. Subsequently, they invited the most skilled workers of these companies to set up shop in Florence and train local people; the import of know-how in tapestry art and production from the Low Countries is a good example that illustrates the case well. In addition, the Medicis established and funded artistic workshops under their supervision, creating a state enterprise for the production of objects of great luxury and prestige. Many of the invited local and foreign artisans and artists were put to work in the Palazzo Vecchio, the official residence of Cosimo I; later also in the Uffizi as well as the Palazzo Piti and its gardens. The state enterprise they established included the production of goods that required precision work and demanded familiarity with the latest insights in science and technology. The goods produced included clocks, quadrants, compasses, armillary spheres, terrestrial globes, astronomical and geographical maps, glass instruments and lenses.

Since the Medici were always either de-jure or de-facto the political power in Florence throughout most of the 15th century, one could talk of an entrepreneurial government that emphasized the importance of public-private partnerships and private initiative to enable people to make optimal use of their skills and thus lift themselves and their families out of poverty.

This investment in human capital, the promotion of entrepreneurship and the relative political stability helped prepare Florence for the regionalization of the western

Mediterranean economy thanks to the transport revolution (improvement in sailing and navigational techniques) that emerged in the second half of the 14th century. Even though Florence could not become a maritime power itself, it managed to take full advantage of the new opportunities in trade because its people had unique skills and sold products that were highly in demand.

At the same time, Florence was less prone to being heavily affected by external economic shocks, due to the middle class that emerged thanks to the institutional reforms. This middle class had a decent amount of savings (as documented in the increase of bank deposits at that time) and, whenever the export-oriented markets in Florence lost an important market to rivals, the shock could always be absorbed by resorting to the home market and selling more goods in the region.

Dealing with the church

There is no doubt that poverty, social inequality and criminal violence was nevertheless widespread in the city of medieval Florence. Such phenomena always appear once people get more political and economic rights because many tend to abuse them. Moreover, the city was almost permanently at war with other Italian states and other European powers. Yet, all these wars mostly occurred outside the city walls and apparently did not affect the economy within the city in any lastingly negative way, judging from all the major public works that were funded and accomplished in the 15th century.

The business spirit of Florence had, however, a potential opponent in the Church, which warned about the divine punishment for people who dedicate their lives to making profits. Avarice and usury were regarded as especially major evils and in his divine comedy Dante designed especially painful hells for those who practiced them. Yet, the churchmen in Florence were also well aware of the importance of wealth generation for the flourishing of the city as well as for the support for their social institutions. Moreover, they were also aware that it was more difficult to impose religious oppression on well-traveled people with cosmopolitan values. For the rich who were still concerned about afterlife, the church went a long way toward placating their anxiety via confession and the institution of the purgatory. The Church put instruments at the entrepreneur's disposal such as commemorative masses, private chapels and donations to welfare and religious institutions and the patronage of religious art. In terms of usury, an accommodation of theology to practice was made. Florentines, at whatever cost to their conscience, found ways to handily circumvent these restrictions – or made gestures of atonement in their final testaments. Churchmen in return tried to adjust doctrine to reality. They were even paying low interest rates for loans to build Monte di Pieta, a religious institution designed to aid less fortunate people by providing an alternative to the Jewish money lending system.

Supportive Humanist Ethics

Humanists who had close relations with the entrepreneurial class in Florence further pointed out the positive aspects of entrepreneurship for society. They used the classical concept of magnificence to explain that wealth is good for the individual because it liberates him from constraints and in turn allows him to liberate others from their constraints. Moreover, it was argued, it helps one to become virtuous and to participate in civic life.

Interestingly, humanists were much less concerned with the generation of wealth than with the question of how to spend it wisely.

Good governance facilitated by the people of Florence

In Renaissance Florence, the underprivileged seem to just have followed the evident principals of political economy using their informal networks to lobby for rules that were more conducive to entrepreneurship and empowerment. In turn, the privileged were not just rent-seekers living from the taxes and land rent as was usual in this period of feudalism, but were engaged themselves in international commerce and business development. In other words, the political and economic institutions were largely a bottom-up process designed by people who did not regard politics as a non-zero game. These institutions largely allowed for long-term business planning by increasing predictability and reducing uncertainty.

No one had the pretension at that time to know how economic development itself should be designed by means of a wise social planner. The adoption of top-down planning emerged only with the rise of economic theory in the 19th century and came to its climax with the use of linear programming and general equilibrium models to predict future economic development. Many of its models were first developed in communist countries but then also adopted in neoclassical welfare economics. All are based on a purely rational and materialist view of the human being. While it was naïvely assumed in communism that this individualism would be moderated by the common long-term goal of a utopic society characterized by ubiquitous social equality, modern economic theory started from the equally naïve assumption that every individual grows up in a social vacuum and with only a sort of innate knowledge or rationality about how to pursue his or her material self-interest.

In the capitalist society of Florence, the term "individualism" would have not been comprehended (Goldthwaite, 2009). No one could imagine himself to be someone outside his family, his guild or his social network. There was still a full awareness in the commune that things can only be achieved together. It was this corporatist spirit that made Florence competitive and humanist alike. This great period in human history allowed people to explore the world and take advantage of its opportunities because they had a solid base of social support. It also enabled them to become risk-takers and innovators. Their achievements not only benefited them as individuals but also their commune and, ultimately, all the subsequent generations worldwide because their activities are found in the great archives they left behind documenting all the disciplined and formal activities of this time.

Concluding Remarks

The current economic downturn will have an impact on future academic research and training, especially in the fields of ethics and economics. These two academic disciplines have enjoyed high prestige up to now due to the general belief that they could provide reasonable guidance in a fast changing and increasingly interdependent and uncertain world. Yet, guidance in such a complex world should not be based on disciplines that have largely failed to renew themselves over the past two decades in the face of the increasing complexity that results from accelerated economic and technological change. During the Cold War period, both disciplines focused on an exclusively deductive approach to investigate how the human being does act in life and how it should act. For that purpose they started from given principles and premises that reduced the human being to a conscious and reason-based being that uses pure rationality to either serve himself (economics) or serve others by considering other-regarding preferences that are not related to self-interest (ethics). As illustrated in this article, the global food and financial crises serve to show that humans badly fit either such narrow and static definition.

The inductive approach applied in laboratory experiments has revealed that, for the most part, bodily emotion trumps "pure" reason in every human decision. Moreover the human being is an active searcher for patterns in life that provide orientation and meaning and guidance in the pursuit of self-interest. This search is an adaptive process of trial and error.

In the course of the evolution of the mind, the individual has realized the need to learn how to act in two parallel frameworks. On the one hand, he needs to respect the informal rules of his community, which provides him with a solid foundation of mutual trust and support. His embeddedness in this community enables him to be sufficiently equipped to pursue his self-interest in the larger world outside the community. But even in this larger world he needs to respect people and understand their points of view if he wants to do business (or interact with others in any way). Therefore formal rules are jointly designed to make interactions and transactions beyond the personal community predictable and mutually beneficial. The pursuit of self-interest in this bottom-up approach must be seen in the context of both, economic and social needs. The human being needs a solid community network on which one can fall back if things go wrong in the business practice beyond the community.

Due to a cult of individualism in recent decades, many people feel that they need to separate from the original community (including parents and siblings) in order to find their true selves. The belief in the true self is again grounded in an idealist view of human nature therefore related to a world view shaped by classical economics and de-ontological ethics.

In Renaissance Florence, the community in which people grew up was not regarded as a liability but perceived as an asset in the struggle for personal flourishing. Even though the Renaissance period was also characterized by increasing economic globalization, countless wars and the spread of communicable diseases, people did not despair but instead became more creative in their efforts to find solutions to their economic and political problems. Since economic globalization is always accompanied by more intense interethnic collaboration it also goes along with an extension of moral imagination, which includes the insight that treating others with respect and moral sensitivity is in the self-interest.

Even though macroeconomic theories and ethics outside theology did not exist in Renaissance Florence, economic and social policies were firmly grounded in the anthropological observation that human action is and always has been, guided by emotion as much as reason. The social and economic institutions at that time were not planned from above but grew from below. As a consequence, public policy was based on the proper understanding of the real human nature rather than derived from an abstract ideal.

Ethics and economics are the two academic disciplines that could learn most from the Renaissance Florence experience in their efforts to adjust to the fact that their preferred theories were not helpful, neither in predicting nor in addressing the food and the financial crises of 2008.

In practice, they would not even have to return to the transformational period of the 15th century, but could learn much from the policies practiced in emerging economies. They have managed the crises better than the industrialized world and thus seem to be fitter alternatives to cope with global business cycles and problems of rural poverty. Renaissance Florence and the most successful developing countries have in common that their policies are firmly rooted in observation and experience rather than abstract theory. Moreover, each can be seen to view economic human development as an evolutionary process rather than a comparative static concept and their resulting political strategies as conducive to entrepreneurship, public-private partnerships and integration of the global knowledge economy.

References

AERNI, P. (2007) "Exploring the linkages of commerce, higher education and human development: A historical review", ATDF Journal 4(2), p. 35-48.

AERNI, P. (2008) "A New Approach to Deal with the Global Food Crisis", *ATDF Journal* 5(1/2), p. 16-32.

AERNI, P. (2009) "What is Sustainable Agriculture? Empirical Evidence of Diverging Views in Switzerland and New Zealand", *Ecological Economics* 65, p. 1872-1882.

AKERLOF, G. A. and SHILLER, R. J. (2009) *Animal Spirits*, Princeton, Princeton University Press.

BEINHOCKER, E. (2006) *The Origin of Wealth*, Boston, Harvard Business School Press.

CAMERER, C. F. (2006) "When does 'economic man' dominate social behavior?" *Science* 311, p. 47-52.

COCHRANE, W. W. (1979) *The Development of American Agriculture: A Historical Analysis*, Minneapolis, University of Minneapolis Press.

DEWEY, J. (1910) *How We Think*, Boston, D.C. Heath.

GOLDTHWAITE, R. A. (2009) *The Economy of Renaissance Florence*, Baltimore, John Hopkins.

GRÜN, K.-J. (2009) *Angst*, Berlin, Aufbau Verlag.

HAIDT, J. (2007) "The New Synthesis in Moral Psychology", *Science* 316, p. 998-1002.

HERRING, R. (2009) "Peristent Narratives: Why is the 'Failure of Bt Cotton in India' Story Still with Us?", *AgBioForum* 12 (1), p. 14-22.

MASTERS, W. (2008) "Beyond the Food Crisis: Trade, Aid and Innovation in Agriculture", *ATDF Journal* 5(1/2), p. 3-13.

REDFORD, D. B. (Ed) (2002) *The Ancient Gods Speak: A Guide to Egyptian Religion*, Oxford, Oxford University Press.

ROMER, P. (1994) "New goods, old theory, and the welfare costs of trade restrictions", *Journal of Development Economics* 43, p. 5-38.

ROSSET, P. (2009) "Agrofuels, Food Sovereignty, and the Contemporary Food Crisis", *Bulletin of Science, Technology and Society* 29(3), p. 189-193.

SOLOW, R. S. (1957) "Technical Change and the Aggregate Production Function", *Review of Economics and Statistics* 39, p. 312-20.

SPINOZA, B. (1677) *Ethik in geometrischer Ordnung dargestellt*, Hamburg, Felix Meiner Verlag (veröffentlicht 1966).

STRIEDTER, G. S. (2004) *Principles of Brain Evolution*, Sunderland, MA, Sinauer Associates.

THE ECONOMIST (2009a) "World Food Prices: Whatever happened to the food crisis", *The Economist*, July 2nd, 2009, p. 68-71.

The Economist (2009b) "The dragon in the backyard", *The Economist*, August 15th, 2009, p. 17-19.

The Economist (2009c) "The other-wordly philosophers. Briefing the state of economics", *The Economist*, July 18th, 2009, p. 68-71.

The World Bank (2007) World development report 2008: Agriculture for Development, Washington DC, The World Bank.

Von Hibbel, E. (2006) *Democratizing Innovation*, Cambridge, MA., MIT Press.

Weil, S (1986) *An Anthology edited and introduced by Siân Miles*, New York, Grove Press.

Wright, R. (2009) *The Evolution of God*, Boston, Little, Brown & Company.

1.3 Collective Intelligence and Business Enterprise 2.0

by Florence Devouard[1]

For countries in the vanguard of the world economy, the balance between knowledge and resources has shifted so far towards the former that knowledge has become perhaps the most important factor determining the standard of living – more than land, than tools, than labor. Today's most technologically advanced economies are truly knowledge-based.

World Development Report, 1999

After World War II, the demand for most goods was higher than the supply. Companies needed to develop new technologies to be able to meet the demand and produce enough goods. Whilst constrained to respect the laws and principles of social justice, this production drive was essentially led by the market. As the supply levels moved upward, active marketing came into action to stimulate the market. Later, in the eighties, in a context of market saturation, quality became the lead. The economic environment shifted again in the mid-nineties.

Today, as supply has clearly exceeded the demand for many products and services, innovation is generally perceived as one of the fundamental sources to increase wealth in the economy. Within this framework, innovation can be regarded as something which increases value, for the customer and/or for the producer.

1 Florence Devouard is a founding member of Wikimedia France. She served as one of the elected wikimedia community representatives to the Board of the Wikimedia Foundation* during four years and as Chair of that Board (2006-2008). She is now a member of the Advisory Board of the Wikimedia Foundation and acts as an independent consultant on issues related to knowledge-management collaboration tools and online communities, as well as a speaker on collaborative and social-media topics.

Innovation begins with creative ideas. It can be defined as the successful implementation of creative ideas within an organization. In this view, a body of knowledge and a "spirit" of creativity at individual and team levels are the prerequisite raw starting materials for innovation to occur. But innovation is also a management process that requires specific tools and rules to give it direction. The shift to knowledge as the primary source of business value means the new economy is led by those who manage knowledge effectively – ie those who create, find, and combine knowledge into new products and services faster than their competitors.

In this essay I have attempted to provide a brief résumé of the changing world of "knowledge" as the word applies in the framework of commercial enterprise. The Internet has brought with it an unprecedented rapid evolution of new language and of new concepts associated with old language. It is challenging to discuss the latest tools and methodologies available to our knowledge society without employing a number of terms that some might qualify as jargon. Where I have deemed it to be useful I have indicated links to Internet sources of information that should help clarify the particular sense implicit in my use of certain language.

Explicit and Tacit Knowledge

The knowledge-based economy obliges us to consider continuous learning and knowledge acquisition as critical elements for a company's success. All knowledge is not the same though. One can divide it into two categories; *explicit knowledge* (e.g., patents, formulas, methods) which can be written down and is often acquired through formal education and training and *tacit knowledge* gained from experience (e.g., relationships, values), which is embedded into an organization's practices. *Tacit knowledge* cannot be easily written down or codified but is as just important as the formal, codified, and perceptibly structured *explicit knowledge.* Another type of simplistic classification makes it possible to distinguish broadly understood different types of knowledge, (even if different individuals cannot reach universal agreement on what things qualify as facts and what things as beliefs). *Know-what* is knowledge about facts. *Know-why* is the knowledge about the natural world, society, and the human mind. *Know-who* refers to the world of social relations and is knowledge of who knows what and who can do what. *Know-where* and *know-when* are becoming increasingly important in a flexible and dynamic economy. *Know-how* refers to skills, the ability to do things on a practical level. Each of those Know-criteria is internalized differently in companies.

Economic growth is driven by the accumulation of knowledge

Business performance in the knowledge society is no longer just about producing and interpreting facts, but also about mobilizing the tacit knowledge and collective intelligence of its stakeholders. For this to happen, the company needs to build a *learning capacity* within its organization. Learning to acquire tacit knowledge and experience must be a permanently ongoing process. Such knowledge is unlikely to be available in a structured format and so cannot be acquired by formal education and training. Instead it requires a continuous cycle of discovery, dissemination, and the emergence of shared understandings. Tacit knowledge is a greater source of sustainable competitive advantage as it cannot be easily replicated by others, however it also can be lost much more easily than explicit knowledge. The deciding factor between success and failure in business often comes down to how well available explicit knowledge is located, leveraged, and blended with internally generated and collected tacit knowledge.

In a globalized world-economy, companies might also gain a competitive advantage when new products and services are imagined, then developed, in a more responsible, more thoughtful way, compatible with the vision of a desirable future that appeals to their stakeholders. Companies may also consider changing their relationship with their customers so that the latter become more involved in the creation and production processes. What would be the purpose? In short, value creation depends mainly on ideas that are in people's heads. Whether these people are customers, employees, partners, or suppliers, ideas cannot be taken by force or by ruse. Thus it is in the interest of companies to provide an enabling environment that can help ideas to emerge and flourish. I would argue that it is in their enlightened self-interest for all organizations, institutions or companies to invest in discovering and teaching people within them to better communicate, cooperate and collaborate.

Defining collective intelligence

Pierre Lévy and others have written eloquently and at length on the prospects of mankind's emerging world in cyberspace for developing the notion of collective intelligence (CI) (Lévy, 1994). Depending on the context, the term also applies to a wide variety of forms of collaborative processes observed in bacteria, animals, humans and computer networks. According to Jean-François Noubel, founder of a web-science proficient Internet community[2], with the self-imposed mission to encourage and disseminate research in collective intelligence (CI), The term CI can be defined as the ability of a group of people to work together to formulate its own future in a complex system.

The human species has social skills based on collaboration and mutual support. It is commonly held that such skills are maximized in small groups (10-20 people maximum) when both the individual and collective benefit is higher than would have been obtained if each had remained isolated. The word *emergence* is used, to refer to what occurs when a new level of complexity emerges from the collective (commonly expressed as "the whole is greater than the sum of its parts").

At the heart of the functioning of these small groups, lie the following pillars:

- A convergence of interest (at the individual and collective level);
- A social agreement (or contract). This contract may be implied or explicit, objective or subjective. The agreement comprises the goals, the values, and the rules of the group. It must not only be acknowledged but also approved by each member of the group (or even better, collectively created);

---

2 http://www.thetransitioner.org

- A holoptic space. This space is characterized by a spatial proximity that gives each member of the group a full and constantly updated view of the other members, their activity, as well as of the emerging level. Holoptism is opposed to panoptism, the concept of allowing a unique person to observe others without the others being able to tell if they are being watched and without others being able to observe each other;
- A polymorphic social structure, where each member takes the lead as needed. Relationships between the participants evolve constantly according to circumstances, expertise, perception of each member's ability to fulfill the tasks or follow the rules defined by the social contract;
- An economy based on gift. In the traditional competition-based economy, the individual gets a reward in exchange of an effort. In the gift economy, the individual will give first, and will get something back only once (if ever) the group has gained in wealth. Most of our social life is based on the gift economy;
- The existence of a learning organization, where both individuals and the collective can learn from the experience of the parts. The group has an ability for taking into account errors made by each member of the group and by the group itself as well as an ability to embrace both complexity and the unexpected (referred to as *serendipity*, the effect by which one accidentally discovers something fortunate, especially while looking for something else entirely unrelated).

These characteristics define what Noubel calls the original collective intelligence, as met in ancient tribes or in present-day sport teams.

The activities and success of these small groups meet two natural limits:

- The number of participants: only a small number of people can interact efficiently;
- Spatial proximity: members should be in the same physical environment so that everyone can grasp the entirety of what is happening and adapt his behavior accordingly.

To circumvent these limits and make it possible to deal with larger groups (companies, states, religious organizations, etc…), other approaches have been implemented. Today, most medium and large organizations have a similar infrastructure based on:

- Authority (which creates an asymmetrical information access, based on command and control);
- Division of labor (specialized access to information);
- A strictly defined organization chart (individuals handle tasks given by their manager rather than according to their ability or to the collective need);
- Often proprietary norms and standards (not transferable or usable by other organizations);
- A fixed monetary system based on scarcity.

Until recently, this form of social architecture was the only system available to guide and organize complex groups. Its use is particularly successful for mass production that can benefit from economies of scale. It is sometimes referred to as *pyramidal intelligence*.

This type of pyramidal organization remains effective as long as the environment is stable, slow and linear, but quickly becomes vulnerable and inefficient in highly dynamic or poorly predictable situations.

Another type of collective intelligence is described as *swarm intelligence*. Swarming has sometimes been defined as a type of artificial intelligence based on the collective behavior of decentralized, self-organized systems. We can observe it in insect societies (e.g., bees, ants) and is generally associated to stigmergy, a mechanism of spontaneous, indirect coordination between agents or actions, where the trace left in the environment by an action stimulates the performance of a subsequent action, by the same or a different agent (e.g. typically the pheromones of ants). A swarm is a complex, apparently intelligent structure, without need for any planning, control, or even communication between the agents. As such it supports efficient collaboration between agents, who may lack even awareness of each other.

Whilst a form of self-organization, swarming mostly lack the holoptism characteristic of Noubel's original collective intelligence. Swarm intelligence is nevertheless used in human society in the context of forecasting problems, e.g. for *prediction markets*. It has been the object of much of the work of Don Tapscott and Anthony Williams (Tapscott, 2006).

Toward a global collective intelligence

Today, most companies face the challenge of a globalized business world, complex, unpredictable, and in some sectors, rapidly evolving and chaotic. Businesses are also faced with permanent conflicts of interest between profitability and sustainability, secrecy and transparency, collaboration and competition, or information sharing and information protection. The pyramidal system may not be the best solution in many cases. Moving toward a different organizational structure might facilitate the emergence of a global collective intelligence system, promoting creativity, flexibility, responsiveness, smarter decision-making, innovation and even perhaps … social friendliness!

However, this shift will only be possible if the principles of original CI are respected.

In addition to the principles of original CI listed above (i.e., social contract, holoptism, polymorphic social structure etc.), the new system would probably have to accommodate new features, including:

- An alternative to the present monetary system as an intermediate tool to coordinate the reciprocal exchange of goods and services, and which would be no longer based on scarcity but on adequacy and availability;
- Greater use of open norms and open standards, essential to organize the cohesion, permeability, and interoperability in large groups. Those standards and processes should be emerging bottom-up rather than top-down;
- A new information system, to organize and optimize the space shared by individuals, to bring people together, to host knowledge, etc. The new system should essentially extend holoptic space by compensating some of the limitations imposed by physical proximity with a virtual proximity (*virtual* meaning *online* rather than *not real*).

Does this vision seem highly impracticable? Not so much. Consider:

The new information exchange system is already there: the Internet.

Interoperability, a property referring to the ability of diverse systems and organizations to work together (to inter-operate) is often used in a technical systems-engineering sense, but it should rather be extended to take account of social, political, and organizational factors that impact the entire system performance. In the early 21st century, interoperability is a matter that generates much debate (e.g., EC competition law cases involving Microsoft and *the operability remedy*).

The new monetary system is coming (it need not necessarily be understood as our traditional idea of currency, but rather as any intermediate tool which facilitates equitable, complex exchange of goods and services). A new reciprocal reward system is already emerging on the Internet, where many products and services (e.g., pictures, videos, music, cooking recipes) are no longer exchanged against a financial compensation. The Internet can be seen as developing itself on various business models amongst which two are quite dominant:

The gift economy occurs where valuable goods and services are regularly given without any explicit agreement for immediate or future rewards. However, the participation of each member increases the wealth of the collective. A prime example is Wikipedia[3], the online free encyclopedia built by Internet users: the user offers his knowledge and benefits back from the collective knowledge.

The attention economy is an approach to the management of information that treats human attention as a scarce commodity. Internet users have access to huge amounts of information. Confronted with *infobesity*, their attention is rare and volatile. As such the human attention becomes a currency and is largely responsible for the success of targeted advertising such as Google Adwords. Other initiatives have led to non-profit exchange networks (e.g., LETS, Local Exchange Trading Systems) or to the creation of virtual currencies (*open-money*), which are not based on tangible and material wealth, but on trust and knowledge that people have to offer. An interesting example is the recommendation systems developed on popular virtual social networks (such as LinkedIn).

In short, an evolutionary step up from the original collective intelligence, which can exist only in small groups, to a global collective intelligence, has become imaginable!

Internet power and enterprises

Let's consider some figures that stimulate the imagination. Number of people connected to the Internet in 2008: 1 billion. Number of people with mobile phones in 2008: more than 3 billions (many with Internet access). According to theories of network space developed by sociologists and mathematicians, there are no more than *six degrees of interconnection* between all of us on the planet[4]. Has this number reached some finite limit? Will it grow or shrink as the world's population rises, or if such technologies had universal access?

---

3 Wikipedia, notable example of a project implementing Global Collective Intelligence principles (Paumier and Devouard, 2009).

4 http://en.wikipedia.org/wiki/Six_degrees_of_separation

I distinguish several reasons why most businesses cannot ignore the major changes that have come about due to the rise of Internet;

Most companies can no longer afford to consider their market is solely in their city, or even in their country. Customers, suppliers, partners and … competitors … should be considered not only at the national level, but at the global level.

To operate efficiently, workers need access to information, regardless of their geographical location and of whether they are traveling or based in one location. This information must be complete, accurate, factual, up-to-date, and transferable.

The evolution towards a single global economy also obliges an increasing reliance on external partners to support non-specialized activities, whilst a company focuses on its core business, its real expertise. With what implications? More offshoring, outsourcing, telecommuting, often with insufficient communication and data sharing between workers due to firewalls, access restrictions, confidentiality requirements can result in the loss of the holoptic space necessary for rich collaboration.

To be fully efficient, the company needs to establish channels of cooperation and communication not only internally between co-workers, but externally as well, between workers, partners, consultants, subcontractors, customers… Internet tools are a great way to facilitate access to information, cooperation and collaboration.

Demographics; the new generation reaching adulthood (sometimes called *Generation Y*, born between 1978 and 1995) has never known a world without Google, without a mobile phone or ipod. This generation has been shaped by the instant communication technologies made possible through the use of Internet. Young adults are largely using communications tools such as email, text messaging, chats, or social networking websites, and are much more peer-oriented than the people from the previous generations. This generation has values different of those of the previous generation, which can be said to have relied on values of loyalty, security, hierarchy in the professional environment. Generation Y pays more attention to creativity, social connectivity, fun, freedom, instantaneity, and diversity. In sticking to traditional organizational models, companies may completely alienate employees from this new generation.

Another consideration is the trend towards the customer becoming a *prosumer* (a composite term formed by contracting the words producer and consumer). In contrast to the consumer, the prosumer takes an active role in the production process. For economists, the prosumer also has a greater independence from the mainstream economy than the traditional consumer archetype. In both cases, the prosumer gets informed and informs others, frequently through social media.

This rise of *prosumer* power also gave birth to a new term, *crowdsourcing*. *Crowdsourcing* is a neologism for the act of taking a task traditionally performed by an employee or contractor, and outsourcing it to an undefined, generally large group of people or community in the form of an open call (InnoCentive[5] is one current example of *crowdsourcing*).

What are the implications of these new trends? Customers increasingly request tailored products and services from companies. They have the ability to look for information and seek advice from other customers on the Internet. They want to be able to publicly

5 http://en.wikipedia.org/wiki/InnoCentive

comment, criticize, support or reject products and services. They request their wishes to be taken into account.

When the company has a presence on Internet (also known as *digital identity*) and creates the right channels for communication, smart discussions between the company and its customers may occur, resulting in additional customer satisfaction; otherwise, valuable marketing data will be lost. Through social media, companies may engage communities to support the creation, the development, and the management of a company mark: this is known as *community marketing*. Bloggers in particular are increasingly used by companies to help raise community interest in a product, or help collect feedback on new products.

Finally, the image of a company or product is quite difficult to control when consumers have the power to share a criticism with millions of other consumers within a few minutes or hours. Even worse, the reputation of a company might be damaged without the company even realizing.

Web 2.0 refers to the second generation of web development and web design that facilitates information sharing and collaboration on the web. It may also be called the participatory web, to refer to the ability it offers to all Internet users to interact with each other. By extension, *Enterprise 2.0* may refer to the introduction and implementation within an enterprise of web 2.0 technologies but it also refers to the social and organizational evolution of a company toward more openness, more collaboration, and more sharing.

Simply described, the 2.0 company is what results from the implementation of the principles governing the collective intelligence at the enterprise level.

Collaboration versus cooperation versus coordination

An additional clarification could be made at this point. The term *collaboration* is widely used (e.g., collaboration tools, collaborative environments, collaboration platforms) to the point where its usage seems almost indistinguishable from the terms *cooperation* and *coordination*. But for me, the three terms do not really carry the same meaning in the business framework.

The goal of coordination is to harmonize tasks, roles, and schedules in simple environments and systems, so as to avoid overlapping in individual assigned works. It requires shared goals and understanding of what needs to be done by whom, which may be supported in particular by project management tools. The role of the manager is mostly to identify the tasks, distribute the tasks and check their completion.

Cooperation is rather meant for solving problems in complex environments and systems, and aims at getting mutual benefits by sharing or generally partitioning the work to be done (thus saving on time and costs). It requires not only shared goals, but also mutual trust and acknowledgement of mutual benefit of working together. The enablers of cooperation are frequent discussions, knowledge sharing, or clear role-definitions.

In a strict sense, collaboration is meant to enable the emergence of understanding and realization of shared visions in complex environments or systems. Compared to cooperation, there is more chance to meet innovative, breakthrough results from the act of collaboration. It requires not only shared goals, trust and respect, but also a sense of commitment, of appropriation, of belonging, open communication and complementary skills and personalities. The primary reason to use the collaborative approach is to achieve results that the participants could not have achieved working alone. Tools for enabling collaboration will be those helping to identify the individuals with specific skills and competencies, open-meetings, brainstorming tools, co-creation tools. The role of the manager of collaboration is mostly to find the right mix of individuals for the team.

Most teamwork is quite naturally a mix of the three elements; coordination, cooperation, and collaboration. But in practice most projects actually involve more coordination and cooperation than collaboration. Even large international scientific projects seem to me to be frequently more about cooperation and coordination than true collaboration. I not mean to imply that empowered collaboration is in any sense better, or more valuable than the other two qualities, but that, depending on the tasks and goals to achieve, the preconditions for success, the resources, and the tools provided should be different and appropriate.

Finally, there are also many cases where attempts at collaboration fail due to the phenomenon of *groupthink*[6], a type of thought exhibited by group members who try to minimize conflict and to reach consensus without critically testing, analyzing, and evaluating ideas. In the case of *swarm intelligence*, when the goal is to use the *wisdom of crowds*[7] for its predictive value, it is critically important that members of the group be as independent from one another as possible. However, in the case of a global collective intelligence network, individuals must be interdependent on one another and must interact and bond. It is critically important to make sure the individual creativity, the uniqueness, and the independent thinking is not lost in the group due to groupthinking tendencies. Groupthinking tends to appear in situations of directive leadership (e.g. Germany in the Hitlerian era), homogeneity of the group members, and isolation of the group from outside sources of information and analysis. It is also worth observing that many (if not most) internet-based collaborative platforms dealing with brainstorming or co-creation allow pseudonymity or anonymity, thus favoring an unprecedented degree of participation liberated from self or imposed censorship.

Which tools for a 2.0 company?

To set up a 2.0 company, multiples tools are available, of which only a few are listed below. They aim at facilitating content co-creation, brainstorming, searching, informing etc.

The collaborative tools most frequently cited are commonly known as wikis[8], collaborative platforms on which participants can collectively create content or manage projects. The most famous example of a wiki-based website is Wikipedia, the online encyclopedia, but the creation of an encyclopedia is by far not the only use of wikis and various companies have developped wiki software and services adapted to a professional environment.

The concepts of the blog and the chat room were originally developed through motivation to disseminate information to others. They are also a great way to exchange ideas

6 see work by social psychologist Irving Janis (Janis 1972)
7 http://blogs.salon.com/0002007/2004/11/15.html#a952
8 http://en.wikipedia.org/wiki/Wiki

with customers and collect feedback to improve products and services. Micro-blogging, which is becoming rather mainstream in 2009, is a form of multimedia blogging that allows users to send brief text or multimedia updates such as photos or audio clips, and publish them, either to be viewed by anyone or by a restricted group which can be chosen by the user. Blogs, microblogging and chat systems make it possible to rapidly exchange information within a company, without the need of a complex, slow, or clogged communication system (e.g., corporate newsletter or email).

Syndication systems or Rich Site Summaries (RSS) permit an effective monitoring of blogs and websites. The RSS is a family of web feed formats used to publish frequently updated works – such as blogs or news headlines – in a standardized format. Users of RSS can follow the feeds on a specialized software called a feedreader or aggregator, which can be web-based, mobile phone-based or desktop PC-based. Quick and easy to use, the RSS systems allow workers to save huge amounts of time when it comes to keeping up-to-date with specialized information, and provide a means to share information with other workers on a single interface.

There are other tools, such as social bookmarking (a way to store, classify, seek and share favorites), virtual social networks (websites that increase the opportunities for meetings between staff and provide tools to recommend business partners), heuristic mapping (tools for collaborative brainstorming), community portals, shared calendars, videoconferencing tools, document sharing tools, and mash-ups (an integration technique with the ability to mix information from multiple sources using the web browser itself as a point of integration).

The current abundance of technology makes it likely to imagine that many new products and services will enrich the business toolkit in the years to come.

Tools such as wikis, blogs, microblogging systems, commenting systems etc. host some explicit use-generated content. Information fed into these systems has been given knowingly and deliberately.

However, whilst individuals use the Internet, they also produce a lot of tacit information, such as traces when they travel from one page to another, or traces when they click on a link. Internet users are generally not aware of the richness of data contained in traces. This implicit data may be collected, aggregated to show; what's good, what's cheap, what's popular, what has a good or a bad reputation etc. One such example might be the engines providing the top ten of the most visited websites in the world.

Another type of implicit information results from the aggregation of the explicit data provided by the individuals. A typical example is the visual depiction (*tagcloud*) of the user-generated tags that are used to describe the content of web pages.[9]

Implementation: A social challenge more than a technical one

Generally, these tools are simple to implement and their uses require little or no training. Within a large company, the technical department can take care of the set-up and maintenance of such tools, or it may outsource the tasks. More complex multi-tool, integrated

---

9 http://en.wikipedia.org/wiki/Tag_cloud

platforms are also available from various software companies. For smaller companies the tools may be hosted for a very fair price by hosting companies or may even be entirely available *on the cloud. Cloud computing*[10] is one of the new trends that characterize the Internet in 2009. It is a style of computing in which dynamically scalable and often virtualized resources are provided as a service over the Internet. Users of resources *on the cloud* need not have knowledge of, expertise in, or control over the technology infrastructure in the *cloud* that supports them. This new trend is likely to make a big difference in particular for smaller companies, associations, or starts-up with limited technical expertise and funding.

A benefit of these new tools is that many of them are available as free software, thus avoiding additional costs and making it possible for companies to implement any technical changes that might be useful to them.

Free software is software that can be used, studied, and modified without restriction, and which can be copied and redistributed in modified or unmodified form either without restriction or with minimal restriction. Free software is available free of monetary charge in most cases. It has played an important part in the development of the Internet, the World Wide Web and the infrastructure of many dot-com companies. Some excellent proprietary tools have also been developed, sometimes more suitable to the business world. But practically, the freedom of choice between free software and proprietary software and the competition it subsequently generates has been a strong lead to innovation.

However, the *participatory Internet*, or *2.0 Internet* should be understood not just a matter of tools, but rather as a question of attitude. The increase in the use of 2.0 tools is often invoked as proof of a successful transformation of a business organization, in particular when it coincides with a tangible return on investment; But the adoption of blogs, wikis and other tools is not a purpose in itself; The bigger picture is about vision, process and culture.

Some of the cultural and technological barriers to overcome in a company emerge in the following types of questions framed in the traditional top-down approach to management:

- How can I monitor the system to make certain that what staff members are saying and sharing reflects company vision, mission, and values?
- How do I know if I am getting my money's worth out of the investment?
- How do I "manage" the gathering and disseminating of so much information?
- How can I be sure that information is being "tagged" properly for efficient retrieval later?
- How do I protect the integrity of the information from malicious tampering by disgruntled employees?
- How do I control who has access to particular levels of information?
- How can I be certain that the information that is gathered and shared in the company stays in the company?
- What are the legal dangers in saving and sharing so much loosely supervised input?
- How can I be sure I will still be able to make decisions as leader of the company?

10 http://en.wikipedia.org/wiki/Cloud_computing

These are all valid questions. Staying open requires to open the frontiers of the company to find the best expertise. Collaboration requires the setting up of production principles largely based on equality and the ability of each collaborator to speak up and participate. Control is largely based on peer-review. Sustainable creativity requires to share one's ideas rather than to protect them.

Such requirements raise in particular, concerns related to intellectual property, transparency, and pyramidal organizational structure.

Resistance to move toward a 2.0 company usually has less to do with the tools than the corporate culture. Lack of time, lack of motivation, fear of loss of control over the final decision, habit, refusal to share, top-down process rather than bottom-up, are the most frequent barriers to the implementation of collaborative practices. Whether a company leader really will be able to resist the temptation to silence dissent, foster open collaboration and adopt a hand-off attitude is an open question. This resistance is particularly important in countries where authority and power distance are important (e.g., France or Japan) and opens new fields of thinking and applications in management. A particularly interesting field is the one related to *consensus* and the tools and methods to help raise consensus, or more generally, to seek motivation, feedback, and participation from individuals (e.g., mediation methods, voting systems etc.), or simply to reward individuals based on their behavior within a group rather than exclusively on their personal successes.

A significant challenge lies in organizing collected data in such a way that it stays useful for human to human communication, but can also be used for communication between human and machines or between machines. Another frequently mentioned is the one related to management of data privacy.

Yet another concern is related to the company's image. Being *open* does not mean the company is disorganized and without rules. Diversity does not mean losing identity. The identity of the company must be developed, communicated, and preserved.

Some businesses have learned to their detriment that bad reputations are made very quickly on the net. In the new economy, a company cannot completely control its image. The best strategy is probably to take note of public criticism, to adopt a genuine attitude in trying to respond to it, and take the time to develop the company's digital identity in a way that reflects its core values.

The decision to share or protect information poses a serious dilemma. There is a clear conflict of view between the proponents of sharing information and those who are supporters of patents and traditional copyright protection. The most frequent attitude if an interesting innovation is developed, is to protect its commercial value by preventing its free use by other parties (with copyrights, trademarks, patents, design rights, trade secrets). Protection of the commercial value of an innovation often facilitates the return-on-investment (but slows down the dissemination).

However, most importantly, the present legal framework for ensuring protection also prevents other parties to brainstorm and develop improvements of the innovation, which might be more valuable. By default, any creation is protected, and the production, then dissemination, of a derivative needs the agreement of the copyright holder. To challenge the *status quo*, supporters of open source, free software and free content have pushed for the development of various licensing schemes to give everyone the freedom to decide what other people could do with their creation (e.g. granting rights for using, selling, modifying the creation, deciding when the creation is released etc), rather than taking such decisions later, upon request, on a case by case basis.

One example of such a licensing scheme is that proposed by the Creative Commons. The Creative Commons[11], a non profit organization, has released several copyright licenses since 2002. Many of these licenses grant certain "baseline rights", such as the right to distribute the copyrighted work without changes, at no charge. Some of those licenses grant the right to modify a Work, whilst others prevent it. Some grant the right to anyone to make commercial use of a Work, whilst others prevent it. The outcome is a collection of licenses granting various rights upon release of a creation. Any producer of a new Work may select that license most appropriate to his or her situation.

Two of the licenses proposed by the Creative Commons are considered to be *free licenses* by the Free Software Foundation (FSF)[12]: the *attribution share alike license* (*cc-by-sa*) and the *attribution license* (*cc-by*), though some argue that only the former is truely a free license. The *cc-by-sa* allows anyone to use, modify, distribute and make commercial use of a Work. This is the license used by Wikipedia and more generally the one chosen by creators who want to maximize participation to and dissemination of the Work produced. Licensing schemes such as the ones proposed by Creative Commons bring new opportunities for wide-scale collaboration and new business models. In 2009, Creative Commons licenses are available in 43 different jurisdictions worldwide.

However, various problem issues still apply to the use of free licenses.

Most Internet users (and many companies) have little or no idea of what *intellectual property* means and implies. They may publish content or use content without understanding what their obligations and rights are.

Internet users are becoming increasingly aware of the concept of licenses and a significant fraction now tag their work accordingly to their understanding of the situation. However the FSF notes that many creators tag their content as "under the Creative Commons license", which leaves all the important questions unanswered since there are different Creative Commons licenses granting entirely different rights[13].

There is some confusion over the word *free* (in free license), which originally refers to the four freedoms defined by the FSF over a piece of software (freedom to study the code, freedom to use the application, freedom to modify the code, freedom to distribute the code), but is often confused with the monetary value (free as in free beer). Whilst most freely licensed Work is indeed free of financial charge, the freedom extends much beyond the price.

The results of the ambiguity are sometimes amusing. In the scientific field for example, although it is in the process of changing, much academic publishing is today still protected by commercial copyright, with the publisher being the holder of the commercial part (which limits dissemination of Work, e.g., can result in sold out books, or the enduring habit of circulating poor quality copies of scientific articles). At the same time,

11 http://creativecommons.org/

12 http://www.fsf.org/

13 http://www.gnu.org/licenses/license-list.html#OtherLicenses

academic research is overtly fed by the entire research field, and articles heavily refer to other research papers (which suggests the only really important parameter to respect is attribution of authorship).

The argument over copyright issues goes beyond the technical considerations currently most visible in mainstream media (e.g., how to prevent teenagers from illegally downloading copyrighted music), and touches many legal, social, and ethical considerations. The basis of current copyright laws[14] was set up well over 100 years ago, in times where copying and sharing content was a difficult and expensive task and where collaborative works rarely involved more than a dozen authors. Meant to protect (more or less depending on the systems) both the commercial part of the Work and the author through the attribution requirement, the traditional copyright system now seems to have met its limits. More embarrassing is the ongoing battle between the proponents of open-source and proprietary models, where all parties push their favorite arguments to support one or the other model without taking into account the global economic and social impact of each solution.

Managing collective intelligence demands a complete shift from traditional management methods and its implementation requires time, conviction, and thorough thinking about the social, legal, and commercial implications.

Amongst the benefits though, a company that continuously learns from its past, assumes perpetual evolution, listens and communicates with the rest of the world and is turned toward the future, can gain a potentially huge competitive advantage in terms of innovation and agility.

References

Janis, I. L. (1972) *Victims of Groupthink*, New York, Houghton Mifflin.

Paumier, G. and Devouard, F. (2009) *Wikipedia – Découvrir, utiliser, contribuer*, PUG.

Lévy, P. (1994) *L'intelligence collective. Pour une anthropologie du cyberespace*, La Découverte, Paris. (English translation 1999, Collective Intelligence: Mankind's Emerging World in Cyberspace, Perseus.)

Tapscott, D. and Williams, A. D. (2006) *Wikinomics: How Mass Collaboration Changes Everything*, Portfolio.

[14] http://en.wikipedia.org/wiki/History_of_copyright_law

1.4 Science and the Social Contract: On the Purposes, Uses and Abuses of Science

by Sarah Chan, John Harris and John Sulston[1]

The relationship between science and society has long been the subject of philosophical, political and ethical consideration. Science, in so far as it has had a place in philosophy, politics and ethics, has traditionally figured in one of three ways. In the philosophy of science it has featured principally as an object of inquiry – the inquiry being mainly as to the logic and nature of scientific method and more broadly the purposes for which scientific discovery is or ought to be applied. In politics and society science has often featured as the source of both destructive inventions (from the stone axe to the atomic bomb) and constructive, socially useful technologies – from early to modern medicine, electronics and telecommunications and, of increasing relevance in the present day, 'green' technology and innovation aimed at sustainable development. The interaction between ethics and science has been on occasion somewhat more adversarial: science is often portrayed as the source of dangerous or transgressive ideas or technologies, or of threatening ways of proving its conclusions in human and animal research, while ethics tends to be invoked primarily as a constraint on science and the activities of science that are deemed to be "unethical".

This view of science and ethics, however, ignores the fact that there are manifold aspects of science the pursuit of which is not only ethically unproblematic but may be a moral imperative. The role of ethics in science is much more than merely preventive: outside the spotlight of 'controversial' or 'ethically challenging' research, ethics ought to be concerned with what science should be doing as well as what it should not be doing, how it ought to be used, in short an ongoing inquiry into the nature and purposes of science itself.

---

1 Sarah Chan is currently a research fellow in bioethics and law and Deputy Director of the Institute for Science, Ethics and Innovation (ISEI) of the University of Manchester. John Harris is Research Director of the ISEI and holds the Lord Alliance Chair of Bioethics in the School of Law, University of Manchester. Sir John Sulston is Chair of ISEI and holds a professorship in Manchester University's faculty of life sciences.

In this paper we examine the relationship between science and society and the role of ethics in mediating this relationship. We propose a view of science as both a sphere of practice that is contextualized within society through a responsive process in which each exerts influence on and shapes the other, and a social institution in itself.

On this view the science-society relationship is a dynamic and essentially moral one, mutually supportive and mutually dependent. The details of this relationship, its moral and its political basis, help to define the obligations of scientists and of citizens and highlight an important dimension of the role of the state in the contemporary world. This in turn has implications for the practice and regulation of science and the control of knowledge and for social policy as well as the moral and political responsibilities of ordinary citizens, scientists and government. These implications extend across diverse questions within science ethics: from the ethics of scientific practice itself and the moral responsibilities of scientists, through the usual subject matter of bioethics such as issues relating to genetic modification and human enhancement, research participation and the obligations of citizens to science, to wide-ranging problems at the end-user stage of science, such as issues of justice in science, access to technology and commercialization of research.

Science and the social contract

It may be instructive to think of the relationship between science and society as reciprocal; a relationship of mutual benefit and support, a sort of social contract from which both parties benefit but in which both sides agree to compromise in the interest both of a successful relationship and in the public interest.

Such a conception of a relationship in which the parties have, if not a common at least a complementary purpose, replaces the older and more destructive conception of a relationship of mutual suspicion: where science is seen as potentially out of control with a need for vigilant watchdogs and close regulation; where society is viewed by science as uncomprehending of its methods and indeed its language and uninterested in its results unless either things go badly wrong or sensational benefits are expected. We are not of course suggesting the abandonment of regulation nor of vigilance on the part of society nor of responsibility and accountability on the part of science. Rather we believe that a new era of mutual dependence and understanding is dawning in which science offers a beacon of hope for the solution of intractable threats, like that of climate change, infectious disease or asteroid strikes and society recognizes and attempts to address the many factors that militate against successful science, including overregulation, underfunding and the jealous protection of intellectual property.

While this account focuses on the role of science as a public institution, we do not mean by this to exclude from our considerations what is sometimes termed 'private science', that is, science that is funded from private monies rather than the public purse. It is misleading to think that public and privately funded science can be treated as entirely separate institutions; even privately funded science builds upon and benefits from the knowledge base of the public institution of science, while public perceptions and attitudes and the consequent relationship of both forms of science to society tend to blur this distinction. We may say, however, that there is a stronger requirement that publicly funded science be in the public interest, as a special duty arising in response to the direct support provided from public funds. In a similar way privately funded science has special responsibilities to its funders, but these do not replace or cancel out the overall imperative for science as an institution to produce public benefit.

What is science for?

What are the purposes of science and how should science be used? Although the answers to these questions are at the foundation of much ethical reasoning about science, they are less often explicitly articulated.

Throughout the history of science, different views of the nature of science have been promulgated, each with its own characteristics and ideals. The internal ideology of science has progressed through several incarnations: from science as "natural philosophy", in which the acquisition of knowledge for its own sake was seen as the ultimate goal of science; the development of industrial science, aimed at useful applications; to the evolution of disciplines such as sociobiology which aim to understand social and cultural phenomena in terms of scientific rationality, as well as investigations in the sociology of science which aim to understand science itself as a cultural phenomenon. Of course within science as a whole and in individual disciplines this has not been a linear progression; all these ideals, sometimes complementary, sometimes conflicting, continue to be present to varying degrees and affect the practice of science.

Science has been characterized as ideally or essentially a value-neutral activity in which the goal of gaining knowledge for its own sake, of satisfying pure curiosity, is the ultimate aim. This conception of science entails that science ought to be motivated by and directed only at truth-seeking, irrespective of consequences or social context. Alternatively, science has been seen primarily in terms of the humanitarian aims of science, to add to human welfare and fulfill the needs of human society; on this view, the drivers of science and the directions research should take are or ought to be dictated by a social-needs agenda. Recent understandings of the nature and role of science recognize both the need for a certain independence of science from the socio-political agenda in order to maintain a level of objectivity that does not compromise truth, and the fact that science does not take place 'in a vacuum', implying a need to be responsive and aware of the social context of knowledge and its applications (Gibbons, 1999; Verhoog, 1981). Both the quest for knowledge and concern for welfare are legitimate motivations for science.

Beneficence and enhancement

If improving the human condition is among the ethically justified aims of science, then how far ought science to go in pursuit of this aim? On the face of this, it is difficult to see why this question should be relevant – why would we set limits on the extent to which science is permitted to achieve its ethical goals? Yet this is exactly what some have argued for in the case of medical and other technologies that have great potential to improve health, increase human capabilities and promote human flourishing, allowing us greater freedom to live the lives we choose. These interventions and their uses are often grouped under the heading of "enhancements" – a term that tends to be used to distinguish these uses from therapies, although such a distinction may be said to be dubiously grounded (Chan, 2008; Chan and Harris, 2007; Harris 2007).

Enhancement technologies have been the focus of much debate within the field of bioethics. Prominent critics of enhancement have argued, amongst other things, that the drive to enhance and to improve ourselves beyond some hypothetical benchmark standard of 'normal' is morally unacceptable, even repugnant, or that it is contrary to or might alter (presumably for the worse, so the argument goes) human nature (Habermas, 2003; Kass, 2003; Sandel, 2004).

Such criticisms, however, ignore the continuous and overwhelmingly beneficial use of enhancement technologies (Harris, 2007) by humans and the fact that evolution itself, although purposeless and random, has in effect continually enhanced us to a stage of development that critics of further enhancement now wish artificially to arrest. Such beneficial enhancement technologies include "synthetic sunshine" in the form of firelight, lamplight and electric light, written language, the teleportation of the human voice through telephone and wireless and so called smart drugs that enhance cognition (Chan and Harris, 2006; Greely *et al.*, 2008).

One account that has been proposed is that enhancement is something that is of benefit to the individual, regardless of and despite how that benefit might be measured against any actual or idealized norms of 'health'. Viewed in this light, enhancements are simply another form of improving human welfare.

This naturally begs the questions of what we mean by "welfare" and what the welfare needs are that science ought to be addressing. Basic and immediate biological needs are obviously of high importance, since satisfying these tends to be a prerequisite for having any additional needs at all. The concept of human welfare should, however, be more broadly understood to include not only those goods necessary to ensure continued biological existence but also those required or desired within our biographical existence – in other words, to explore those activities that give meaning to our lives as persons, to achieve our goals and fulfill our aspirations[2]. This is why, for example, as well as health care, food and clean water, and shelter, we regard education as among the primary goods of human existence (see for example CSTD report 1997).

Of course, how we should prioritize at a community level between these welfare goals, and how we should distribute the benefits and burdens of science, requires us additionally to address considerations of fairness and justice. In the individual context, however, any gains to health or well-being more broadly conceived (be they 'therapies' or 'enhancements') are beneficial to that individual and hence form a legitimate part of one of the goals of science, to 'do good'.

Science and nature

Ethical objections to science are most commonly raised against new technologies, and one type of argument that regularly arises in relation to any form of new technology – be it genetic modification, reproductive cloning or human enhancement – is the objection that it is against nature. However, an important part of the scientific enterprise is the aim

[2] These may of course include the goal of satisfying our scientific curiosity.

of gaining knowledge in the pursuit of understanding nature. Part at least of the reason for seeking this understanding is so that nature can be deliberately shaped, in the cause of improving human lives. This of course happens when natural processes like disease, and natural life forms like viruses and bacteria are frustrated or killed by medical treatments, vaccines, antibiotics etc. The purposes of science and society are thus complementary: truth and understanding contribute both directly and indirectly to human welfare.

Scientific responsibility and moral responsibility

The idea of "scientific responsibility" has gained currency within the scientific community and in the wider public sphere, both as a standard by which to judge the behavior of scientists and as a means of prospectively dictating how scientists ought to behave. In line with this a movement has emerged amongst scientists themselves, to establish their own codes of conduct for the practice of science. The overlap between what is usually considered scientific responsibility and what we might describe as the moral responsibilities of scientists, however, and the foundation for these moral responsibilities, is less often elucidated.

Internal responsibility

The responsibilities of science and scientists have sometimes been characterized as internal or external responsibilities, internal responsibilities being those associated with the practice of science itself, such as the responsibility to perform scientifically valid research, to maintain objectivity and to report results accurately and truthfully, while external responsibilities pertain to science in the wider context of society and applications of knowledge.

Fulfillment of internal responsibilities is necessary first in an operational sense: because it makes science work. If scientists routinely did not act in line with the general principles of scientific responsibility such as integrity, truthfulness and openness, then the institution of science itself would break down. In this sense, to be successful, science needs to be done properly; and therefore in order to sustain the institution of science, scientists are required to go about performing it in the appropriate way.

To whom, then, is this duty to do good science owed? Scientists themselves obviously have an interest in the maintenance of science as an institution, since it provides them with a livelihood, inspiration and source of interest. They have also benefited from access to the knowledge generated by previous scientific research and therefore have a reciprocal duty to support the institution of public science.

It is clear, however, that the practice of good science is also a benefit to the public outside of the institution of science itself; and hence these internal responsibilities can manifest as duties towards the public and individuals of the public, as well as duties towards other scientists.

We may draw a comparison here with other highly regulated areas such as medicine: the harm that would (and sometimes unfortunately does) ensue from doctors "cheating" in the practice of medicine is immediately obvious. Science, too, is an area that requires integrity in its performance. An initial consequence of "bad science", failure of scientific integrity, is the loss of public faith in science. This again is directly detrimental to science, in that the public will not support science if they lack trust in it. It might also be bad for society in a wider sense, if we accept that science as an institution is good for the public and that the mutuality of the science-society relationship is therefore a good thing. Public mistrust of science hampers its flourishing and thereby denies the public both the benefits of applied research and of being a science-doing society.

Harms and benefits of science

It is increasingly recognized that apart from responsibilities relating to the practice of science itself, scientists can and do have responsibilities within the wider community as a result of their position as the knowledge-gatherers – and knowledge-keepers – of society.

Science can do harm in how it is used (harmful applications), how it is presented (miscommunication, misinterpretation of results) – and how in some cases it is not used (because, for example, of restrictions on access).

In some cases, the external responsibilities of scientists to prevent and not to cause harm may be somewhat at odds with the internal responsibilities and scientific values discussed above, for example openness and freedom of communication. The apparent tension between freedom of science and protection of public interest is most clear in cases where the fruits of science may have harmful applications, the so-called "dual-use" dilemma (Ehni, 2008). A paradigm case in this regard is the basic research into nuclear physics in the first half of last century that led to the development of the atomic bomb. More recently, microbiological research has given rise to deeper understandings of disease transmission, epidemiology and pathology that could potentially save lives as well as extending the body of basic knowledge; but it has also produced the potential for the creation of new bioweapons.

There is obviously a public interest in preventing such data from being made available and thus potentially used for harmful purposes. Aside from the many impracticalities of suppressing such research or the resultant data, however, the very idea of doing so runs counter to the internal "truth" of science. The suggestion that the data could be published in part only, with crucial steps withheld, also goes against the grain of established scientific norms and values such as openness, replicability and verifiability of research. It seems impossible to demand that scientists refrain from carrying out experiments that may be of immense scientific (and indeed applied) value merely because there exists the possibility that the work might be used for harm rather than good.

Yet by their involvement in such research, scientists do have particular responsibilities regarding the knowledge that it generates. For a start, they are uniquely positioned by virtue of their specialized knowledge to foresee possible harms that might result from their work, and therefore have a particular obligation to act to prevent them.

Scientific (mis)communication

As well as harms resulting from undesired or unforeseen applications of research, science has the potential to cause harm in the social context if research findings can be interpreted or misinterpreted in a socially destabilizing manner. Scientists have the potential to contribute to this through deliberate, reckless or negligent miscommunication about their research and its implications.

Many recent and historical instances of this have arisen in the context of human genetics. The history of early human genetic research and its contribution to the eugenics movement of the early twentieth century is one such example that demonstrates how scientists can exert a negative influence on society through the careless or even deliberate mistranslation of scientific results into inappropriate social policy (Beckwith, 1993; Beckwith and Huang, 2005) – inappropriate because based on inaccurate science as well as ignoring the wider social context.

Since the advent of molecular genetics and the sequencing and analysis of an increasing proportion of the human genome, the early tendencies towards extreme genetic reductionism that are now discounted by many scientists have remained present in the public mind, giving rise to fears surrounding genetic discrimination and determinism and popular misconceptions regarding behavioral traits, such as criminality, aggressiveness or sexual orientation. One example that has been discussed in an ethical context is that of the Maori "warrior gene" (Wensley and King, 2008): a purported behavioral genetic link that scientists claimed could explain "anti-social", "aggressive", "violent" and "risk-taking" behavior in the indigenous ethnic group. These claims on the part of scientists threatened to reinforce and legitimize prejudices about and against the ethnic group in question, as well as contributing to public misconceptions about genetics, both with potentially damaging social consequences.

Even without "actively engaging with the external domain" in the way described in such cases, however, scientists may still have obligations with respect to how their work is presented and received. They may, among other things, have a duty to foresee possible misinterpretations with social implications and act to forestall these. This is also relevant in terms of the possible outcomes of research as well as its desired outcomes, something that is especially relevant in terms of media and public communication. Scientists must work with media to ensure accurate presentation of scientific findings, for without due care on the part of scientists and reporters alike, over-simplification or speculation is liable to result in unrealistic portrayal of scientific results.

It is worth noting that the public may have a reciprocal duty in this regard, to be well-informed about science and to form opinions and make decisions on the best information available to them. Likewise the media, as another group of actors in this relationship, have the responsibility to seek out accurate information from scientists to support the stories that are presented to the public.

Beyond these possibilities for science to do harm and the responsibilities that scientists have in this regard, not only to refrain from doing harm but also to act to prevent harmful consequences of their research, there are also tremendous possibilities for science to do good and obligations that attach to the practice of science in that regard. Another aspect of moral responsibility as it applies to science is therefore to try to ensure, so far as is possible or practicable, that the knowledge gained from science is used in a beneficial way.

The moral community: *who* is science for?

According to the "social contract" model of science, the relationship between science and society is governed by an underlying and often unwritten agreement that defines the reciprocal rights and obligations of the parties to the contract (Gibbons, 1999). We have so far discussed what the moral responsibilities on the part of science and scientists might encompass. Before we can consider what other obligations might form society's part of the bargain, we must ask: whom do we consider to be members of the moral community? Who are the citizens who are participants in the contract, and for whom should science be used?

Justice and the global community

Practitioners and philosophers of science have often argued that science is or should be a global enterprise, a community that transcends national boundaries and shares common goals and norms despite cross-national political or ideological differences. In practice, local and national discrepancies – in funding, in policy and regulation, and (not to be ignored) in the resources and infrastructure available to support scientific research – may restrict the extent to which science can be considered to be freely practiced on an equal footing worldwide. However the ideal of 'science without borders' remains a part of the scientific ethos, operating in tandem with other ideals such as openness of communication and scientific freedom as well as responsibility (see for example CFRS report 2008).

This notion of the global science community, when taken in conjunction with the social contract model of science that treats all citizens of the community as participants in the contract, has important consequences for analyses of justice in science and innovation.

Concerns over technologies such as genetic modification and enhancement often include the fear that unregulated or insufficiently regulated use of these interventions will exacerbate social inequalities in a way that is inimical to justice. The simplistic form of the justice-against-(enhancement) technology argument runs something like this: emerging technologies will probably (at least to begin with) be expensive and therefore only the wealthy will be able to access them. This is unjust because a) it is unfair that access will not be available to all equally, and/or b) it will increase existing inequity, possibly with consequences for society that are predicted to be of varying disastrousness. For example, it has been suggested that germline genetic enhancement, because it will be passed on to successive generations, if available only to the well-off will risk creating a class of "genetic elite" and an underclass of the unenhanced, entrenching and exacerbating the divide that already exists.

The most simplistic forms of this argument often proceed to assert that, because of the potential for new technologies to cause or exacerbate injustice, they should be disallowed altogether.

As we have observed above, however, one of the purposes of science is to produce benefits for humans and humankind, and the notion of benefit operates at an individual level as well as a communal one; justice is concerned with how those benefits are allocated. The competing demands of justice, then, do not nullify the need and the moral imperative for science to do good; rather they add the additional obligation of making sure that what we do to distribute the benefits of science is just insofar as is possible.

We tend to think of the problematic aspects of justice within a closed social system: for example, cognitive enhancers in a university, or access to genetic enhancements within the population of the UK. But what reasons do we have for limiting our considerations of justice to the local community, or to national level?

Taking an international perspective, it is at once evident that developed-world norms count as enhanced by comparison to developing countries, for example with respect to the standard of health care, technology and even of the goods necessary to participate in social processes.

It is almost inevitable that whenever a new form of technological enhancement is suggested or comes under consideration, the simplistic form of the justice argument will be invoked to show why we ought not to do it. Justice-critics rant over Ritalin and jump up and down over genetic modification. Yet if we follow the simplistic argument over justice-as-enhancement, as outlined above, to show why we should reject and prohibit the use of new enhancement technologies, we will also logically be committed to rejecting and prohibiting many current technologies whose use constitutes enhancement in the global sense. What these justice-critics should be agitated about and agitating for, respectively, is not the possibility that we as a population might be making use of cognitive enhancers to help our students and academics study harder and the need for a ban on such use. They (and we) should instead be concerned with the reality that society in the developed world already enjoys many benefits of science and innovation that constitute enhancements globally defined and that have an immediate impact on global justice; we should pay more attention to the urgent need in the name of justice to make access to all of these more widely available on a global scale.

Humans and persons

In asking whom we consider to be part of the moral community for the purposes of science, we need to address not only geographical but other boundaries, some of which may be appropriate *moral* boundaries while others are not.

Moral philosophers have long been concerned with the question of what confers moral status and what makes creatures proper subjects of moral concern. In this context, the theory of personhood and the concept of the person as a being of intrinsic moral value have developed as a widely-respected account of moral status (Locke, 1964; Harris, 1985; Warren, 1997). Personhood theory views being a person as the threshold for all sorts of moral rights, in virtue of being a creature who can value life for any or all of myriad reasons. Such a creature has an interest in its continued life and in all those things that make that life subjectively of value, and this fact is precisely what makes its life also objectively valuable (Locke, 1964; Harris, 1985; Warren, 1997).

Most humans are persons and so far, most if not all of the persons we currently recognize are humans; but this does not mean that all humans are persons or that all persons must necessarily be humans. The species boundary between humans and other animals, and indeed between humans and other sorts of life forms, is often regarded as one that also delineates the limits of our moral community. But why should this be so? Moral accounts of personhood do not include any criterion of species membership, and the attributes of persons might well be possessed by non-humans as well as humans. Indeed, there is increasing evidence to show that some non-human species may qualify for moral personhood on these grounds. See for example (Patterson and Gordon, 1993; White, 2007).

The implications of this for the moral responsibilities associated with science are potentially wide-ranging. If we recognize non-humans as persons, we also have an obligation to include them in our considerations of moral community, and to regard them as worthy recipients of the benefits of research, not just as research subjects.

This is not to say that personhood is the sole foundation of moral obligations, although it is a significant one, both in the kinds and the scope of the obligations it attracts. We may also have supererogatory obligations to non-persons, although they are not direct participants. Many animals, for example, are unlikely to meet the criteria required for metaphysical and moral personhood, but they are still subjects of moral concern in virtue of the interests they possess and their capacity to be harmed or benefited in view of those interests.

This moral theory of personhood and interests as grounding different levels of moral concern provides us with important parameters for defining the bounds of our moral community and the scope of moral constraints and obligations with respect to science. It identifies those sorts of beings who we ought to regard as the proper beneficiaries of scientific research under the 'social contract' model of science, as well as the interests of other beings that need to be weighed in deciding what is permissible in pursuit of science. Where the benefits of science to persons come at the expense of harm to other relevant interests such as those of animal test subjects, an appropriate ethical balance must be struck between the two.

Such a theory also indicates the categories of being or thing that should be excluded from moral consideration; it explains why, for example, human embryo research is morally justified – embryos are not persons but neither can they be said to have interests which are harmed or benefited by being the subject of research. They are not in any real sense members of our moral community, nor can they be considered participants in the social contract of science. We have no obligations towards them, but we do have obligations to conduct the kinds of research that will or may lead to life-saving, welfare-increasing technologies, as well as contribute to fundamental scientific knowledge.

Nevertheless, human embryo research has remained socially controversial – although it has been pointed out that even if embryos were considered to be persons, it might still be in their interests to support science through participation in embryo research (Harris, 1975; Harris, 2003; Harris, 2004; Persson and Savulescu, 2009; Savulescu 2002; Savulescu and Harris, 2004; Chan and Harris, 2009a). A purely utilitarian approach to embryo research might dictate that a 'survival lottery', in which some embryos are sacrificed for the good of the remainder, is an ethically acceptable and even preferable mode of research participation.

Research participation: supporting science

Fortunately for us and for science, most research participation does not involve entry into a survival lottery, but there are still important questions to be addressed about when research participation is permissible and justified, and when perhaps it might even be morally required.

According to the 'social contract' model of science that we have outlined, research is an activity of benefit to society that requires, in order to sustain it, a certain level of support *from* society. Public support for science may come in a number of ways, including research funding, but another important aspect of research – and particularly medical and clinical research, that may have most immediate benefits in terms of health care – is research participation.

Conventional attitudes towards human research participation have tended to emphasize protection of the participants above all else, perhaps for good historical reason. This has led to a focus on the 'best interests', narrowly understood, of research participants, and an underlying assumption that participation in research is inherently harmful or against best interests unless and until it can be demonstrated to be otherwise.

The view of science as a beneficial social institution in which there is a strong public interest, however, casts a new moral light on research participation. Given the benefits that research is likely to produce, and given the benefits we as a society have already received and are likely to continue to receive, it may be argued not only that research participation is in general a beneficial (and beneficent) enterprise, but that we may also have a moral duty to it (Caplan, 1984; Harris, 2005; Chan and Harris, 2009b).

This account of the science-society relationship also has implications for how we understand research participation as a moral act and one of social utility. The tendency has been to conceptualize participation in science and the benefits it might bring at the level of an individual transaction or contract between researcher and research subject. This is apparent in attitudes towards research participation expressed in policy and regulation. For example, EU policy in the form of the Directive on Clinical Trials[3] imposes special requirements in relation to research on children and on vulnerable individuals, stipulating that where research on children is carried out, it must "relate directly to a clinical condition from which the minor suffers" and be of "some direct benefit for the group of patients"[4]. The Declaration of Helsinki likewise contains provisions that the risks of research for individual participants must be weighed against the "foreseeable benefits to them and to other individuals or communities affected by the condition under investigation"[5].

These provisions, however, and in general the account of research participation as an individual (rather than a social) contract, misunderstand the nature of the moral community and of society in relation to science. Why should the community of concern, those who contribute and those who benefit, be limited to those who suffer a particular condition? This is not a morally relevant feature. Moreover the research is likely also to be of indirect benefit to those outside that particular group, and should properly speaking be the concern of all of us.

Once again, science should be seen as an institution belonging to the entire moral community and participation in it should be viewed in these terms, rather than on an individualistic basis. Within the global paradigm of science and research participation, it is then a matter of justice as to how, within the community, the benefits and burdens of science are distributed. Overall however it is too narrowly construed to say that justice demands that participation be restricted only to those who are likely to benefit; and misdirected to say that we should segregate moral communities on the basis of health or disease status.

3 EU Directive on Clinical Trials, 2001/20/EC.

4 EU Directive on Clinical Trials, 2001/20/EC Article 4 (e).

5 World Medical Association, *Declaration of Helsinki* (2008), Article 18.

Consider for example the case of orphan diseases. If participation is restricted only to those who will benefit directly from research or those who suffer from the condition themselves, this will severely limit the pool of research participants for rare disease research. Lack of participants will lead to less research-taking place into therapies for these diseases. Sufferers would hence carry a double burden: that of having the ill luck to suffer the disease in the first place, and then of being the only ones who can take part in the research needed to alleviate that disease.

Health burdens are a global problem and need to be addressed in the context of the global moral community. Recognizing a broader definition of benefit in relation to research participation – the general benefit that comes from being part of a science-doing society – is an important step towards reconceptualizing the paradigm of research participation, for the benefit of both science and society.

Access to knowledge and the products of science

We have so far characterized science as an activity of benefit to the community, a social institution in respect of which members of the community have obligations, to support and maintain the institution. This account of the nature of science and its role within society requires not only that the obligations of support are met but that the beneficial outcomes are realized in return.

The commercialization of science and the assertion of strong intellectual property rights (IPRs) can often present a problem in this regard.

It is true that there are many possible interpretations of "benefit" – some ways of exploitation benefit certain parties in a for-profit manner, others benefit the public or certain sectors thereof. It is also, however, surely fair to say that the benefit that accrues from saving a life or ameliorating a severely disabling disease outweighs the benefit of making an already well-off person richer. Of course, the mechanics of innovation and the economic trade-off are rarely as simple as that. In general, though, one of the problems with the way in which the fruits of science are disseminated and made available is the profit-driven corporatization of the biotechnology industry, which too often privileges profiteering over life-saving and disease-curing.

These issues have been examined in some depth by a working group centered on The Institute for Science, Ethics and Innovation (iSEI) at The University of Manchester. Their first report, the Manchester Manifesto[6], concludes that the dominant existing model of innovation serves some necessary purposes but is flawed in a number of ways. It restricts access to the products of science, and often to knowledge as well, thus blocking what should be public benefits; it interferes with the internal flow of information in science, thus impeding progress. Reliance on revenue raised on the strength of exclusive rights patents is leading to complex networks of licensing arrangements that are costly to negotiate and are actually hindering innovation in some cases.

When people participate in research but the benefits are not fairly or justly distributed throughout society, this violates the 'social contract'. Such a violation can occur

6 Available at www.isei.manchester.ac.uk/TheManchesterManifesto.pdf

through enforcement of over-protective IPRs. For example, in the Greenberg case (Greenberg v. Miami Children's Hospital Research Institute, 2003), families suffering from Canavan disease were recruited to a genetic study commissioned and part-funded by a patient advocacy group and carried out by a research institute. The aim of the study was in part to develop tests that could be used for genetic diagnosis of the condition; the research was successful in this and the tests were then patented by the researchers – effectively restricting the access of those who had participated in the research, as well as the public at large, to its products. This was a clear case of restrictive IPRs being used to prevent members of the public realizing the benefits from research which they had supported through both funding and participation. Again, however, access to the benefits is owed not just to the specific disease community on the basis of certain individuals' participation, but to all members of global moral community.

How we manage this problem will be crucial in determining whether science does the good that it can do in benefiting society as a whole, or merely benefits a privileged few. A more open and collective approach will be of particular importance for tackling the largest problems facing humanity: threats from such challenges as climate change, energy supply, pandemic disease and asteroids cannot plausibly be dealt with by piecemeal deployment of IP dominated mechanisms. By way of illustration, research on flu virus is being impeded by the implacable insistence of rich countries on their right to claim IP on their analyses, and the understandable reluctance of developing countries to donate specimens under such conditions.

We also need to be aware, however, that simply removing direct restrictions on access may not overcome all barriers to access. To satisfy the requirements of the 'scientific social contract' in some cases demands more than just passive permissiveness: in circumstances where some sectors of the community may be unable to gain access to knowledge and other fruits of science due to lack of adequate access structures, an active approach is required.

This may be particularly relevant with respect to both the scientific community and the general public in developing countries and disadvantaged regions of the world. Not just unrestricted access but active promotion of access and means to ensure adequate delivery of the benefits of innovation is required, while attention to capacity-building and research infrastructure is needed to ensure global participation in innovation.

Ethics in science and innovation

In this paper we have set out a framework for considering science as a social enterprise and an activity of public as well as academic interest. This framework treats science as an essentially moral activity and indicates the rights and responsibilities of actors in this context – parties to the 'social contract' of science. The purpose of ethics in this framework is both to facilitate and to reflect on the relationship between science and society and to assist in determining the directions this interaction might take.

This will also necessitate a broadening of the territory of bioethics and science ethics to encompass cognate disciplines such as law and sociology. This is not to imply, however, that the normative content of science ethics, what we *should* do (morally speaking) in relation to science and its products, is to be determined by reference to evidence as to what we are permitted to do or social attitudes towards science. The purpose of enquiries in these areas in terms of ethics is to inform rather than dictate the agenda for ethical consideration.

While science is of massive intrinsic interest and importance, perhaps even to the extent that it needs no further justification, its impact on society is such that the moral dimensions and the moral significance of science are inescapable. Thus, however well science can get on without ethics and however badly the moral future of mankind would fare without science, both the future of science and the welfare of humankind are more secure when science and ethics act in harmony.

References

Beckwith, J. (1993) "A historical view of social responsibility in genetics," *Bioscience*. 43, p. 327-333.

Beckwith, J. and Huang, F. (2005) "Should we make a fuss? A case for social responsibility in science," *Nat Biotechnol* 23, p. 1479-1480.

Caplan, A. L., "Is There a Duty to Serve as a Subject in Biomedical Research?," in *IRB A review of Human Subjects Research* Vol 6, No 5, September/October, 1984.

Chan, S. (2008) "Humanity 2.0? Enhancement, evolution and the possible futures of humanity," *EMBO Rep* 9 Suppl 1, p. 70-74.

Chan, S. and Harris, J. (2006) "Cognitive regeneration or enhancement: the ethical issues," *Regen Med* 1, p. 361-366.

Chan, S. and Harris, J. (2007) "In support of human enhancement," *Studies in Ethics, Law and Technology* 1.

Chan, S. and Harris, J. (2009a) "Consequentialism without consequences: ethics and embryo research," *Camb Q Healthc Ethics* 19, p. 61-74.

Chan, S. and Harris, J. (2009b) "Free riders and pious sons – why science research remains obligatory," *Bioethics* 23, p. 161-171.

Ehni, H. J. (2008) "Dual use and the ethical responsibility of scientists," *Arch Immunol Ther Exp (Warsz)*. 56, p. 147-152.

Gibbons, M. (1999) "Science's new social contract with society," *Nature* 402, p. 81-84.

Greely, H., Sahakian, B., Harris, J., Kessler, R., Gazzaniga, M., Campbell, P. and Farah, M. (2008) "Towards responsible use of cognitive enhancing drugs by the healthy" *Nature*, Vol. 456, 18/25.

Greenberg v. Miami Children's Hospital Research Institute (2003) Wests Fed Suppl., 264, p. 1064-1078.

Habermas, J. (2003) *The Future of Human Nature*, Cambridge, Polity Press.

Harris, J. (1975) "The survival lottery," *Philosophy* 50, p. 81-87.

Harris, J. (1985) *The Value of Life*, London, Routledge.

Harris, J. (2003) "Stem cells, sex, and procreation," *Camb Q Healthc Ethics* 12, p. 353-371.

Harris, J. (2004) "Sexual reproduction is a survival lottery," *Camb Q Healthc Ethics* 13, p. 75-89.

Harris, J. (2005) "Scientific Research is a Moral Duty" in *Journal of Medical Ethics*, Vol. 31. No. 4, p. 242-248.

Harris, J. (2007) Enhancing Evolution: the ethical case for making better people, Princeton, Princeton University Press.

Locke, J. (1964) *An Essay Concerning Human Understanding*, Book II. Chapter 27, Oxford University Press, Oxford and London.

Kass, L. R. (2003) "Ageless bodies, happy souls: biotechnology and the pursuit of perfection," *New Atlantis*, p. 9-28.

Patterson, F. and Gordon, W. (1993) "The Case for the Personhood of Gorillas," in *The Great Ape Project*, Ed P. Cavalieri and P. Singer, New York, St Martin's Griffin, p. 58-77.

Persson, I. and Savulescu, J. (2009) "Actualizable potential, reproduction and embryo research: bringing embryos into existence for different purposes, or not at all," *Camb Q Healthc Ethics*.

Sandel, M. J. (2004) *The Case Against Perfection*, Harvard University Press.

Savulescu, J. (2002) "The embryonic stem cell lottery and the cannibalization of human beings," *Bioethics* 16, p. 508-529.

Savulescu, J. and Harris, J. (2004) "The creation lottery: final lessons from natural reproduction: why those who accept natural reproduction should accept cloning and other Frankenstein reproductive technologies," *Camb Q Healthc Ethics* 13, p. 90-95.

Verhoog, H. (1981) "The responsibilities of scientists," *Minerva* 19, p. 582-604.

Warren, M. A. (1997) *Moral Status*, Oxford University Press, Oxford.

Wensley, D. and King, M. (2008) "Scientific responsibility for the dissemination and interpretation of genetic research: lessons from the "warrior gene" controversy," *J Med Ethics* 34, p. 507-509.

White, T. I. (2007) *In Defense of Dolphins: The New Moral Frontier*, Oxford, Blackwell Publishing.

CSTD (Commission on Science and Technology for Development) report (1997), *Assault on Poverty: Basic Human Needs, Science and Technology*.

CFRS (Committee on Freedom and Responsibility in the Conduct of Science) report (2008) of the International Council for Science (ICSU) on *Freedom, Responsibility and Universality of Science*.

1.5 Scientific Knowledge Leads to Moral Responsibilities – Case Study Synthetic Biology

by Anna Deplazes-Zemp and Sebastian Leidel[1]

Modern technology exerts an enormous impact on our society and environment, influencing our daily life. Furthermore, in addition to its considerable benefits, it also raises questions of moral responsibility: To what extent can scientists and engineers be held responsible for the consequences of a new technology? How are they supposed to cope with such potential responsibilities?

In our essay, we will discuss these questions with a focus on the responsibilities that may arise from knowledge in synthetic biology.

We will start by giving a short introduction into the meaning of moral responsibility. Next, we will analyze the role of scientific knowledge and give examples of how the responsibility of scientists has been thought of in the past. We will then focus on synthetic biology with a short introduction to the field. Finally, we will discuss which responsibilities related to knowledge arise in this field and how synthetic biologists deal with these issues.

Moral Responsibility

When we speak about moral responsibility of a person, we should not confuse it with other types of responsibility such as causal responsibility or role responsibility. The first refers to somebody, who causes a certain situation such as for example "he was responsible for the noise" the latter is associated with certain tasks, as the postman, who is "responsible for delivering the parcel". These two types of responsibility can be, but are not necessarily linked with a moral responsibility, which identifies the holder of this type of

1 Anna Deplazes-Zemp is a senior research fellow at the Institute of Biomedical Ethics, University of Zurich. Sebastian Leidel is a group leader at the Max Planck Institute for Molecular Biomedicine, Münster.

responsibility as someone who deserves praise or blame for the action she is responsible for (Eshleman, 2008). Particularly, in the German tradition, responsibility is described as a relational term with several elements ("mehrstelliger Relationsbegriff"). This designation indicates that "responsibility" describes the relationship between a *subject* and an *object*, towards an *instance*, based on *a normative criterion*. The *subject* of responsibility is the moral agent, who is held responsible for the respective act. The *object* is the act itself, or the outcome of that act. The *instance* can be a person, an institution, a nation, or even god towards whom the subject is responsible. Finally, the subject is held responsible based on a *normative criterion*, such as a social or religious rule. The details of these concepts have been debated such that the list of elements that describes such a relationship can be expanded or shortened. However, it appears to be clear that different aspects need to be considered and specified in the discussion about moral responsibility (Lenk, 1991; Lenk and Maring, 2001; Ropohl, 1987; Zimmerli, 1987).

Each aspect of the multifaceted concept of moral responsibility raises a set of interesting questions. For example: Which are the relevant and binding criteria for responsibility? Towards which instances are we responsible, and are future generations or nature such instances? What are the exact objects of responsibility? Are scientists responsible only for their discoveries or also applications that emerge from them? However, the question, which appears to be the most controversial, concerns the conditions under which the subject can be hold responsible: Is knowledge a precondition for responsibility? In other words, can a person be responsible for unforeseen consequences of an act? The question about the subject of responsibility is even more difficult to answer if this subject is part of a group. Under these circumstances, we need to consider the concept of "collective responsibility" and ask whether it is possible to hold a group of people responsible or whether a group lacks certain features that are a prerequisite for responsibility (Smiley, 2008).

In our essay we will address only some of these questions. In particular, we will focus on the *object* "knowledge in synthetic biology" and discuss which responsibilities may arise for researchers towards which *instances*.

The role of scientific knowledge

We focus on moral responsibilities related to knowledge in synthetic biology, because we consider such knowledge as something that might render involved subjects blame- or praiseworthy. Therefore, we will discuss the impact of scientific knowledge in general in order to explain this notion.

The dictum "knowledge is power" attributed to Francis Bacon[2], seems to apply well to our society and its technological progress. On the one hand individuals depend strongly on a good education in order to be able to succeed in today's "world". On the other hand, novel technologies based on scientific knowledge[3] are the driving forces for the development of industry, nutrition, warfare, medicine, entertainment, culture and every other

---

2 *Ipsa scientia potestas est.* (knowledge itself is power) (Francis Bacon: in Meditationes sacrae: Bacon, 1963)

3 We are aware of the fact that scientific and technological knowledge are often distinguished. However, for our article we do not make this distinction. When we speak of "scientific knowledge" basic forms of technological knowledge that are acquired by research rather than by trial and error, are therefore included.

aspect of human life in industrialized countries. This high impact of scientific knowledge grants power and leads to responsibilities. This is even more the case since scientific knowledge and its technological application are so sophisticated that specialized experts are required. Therefore, for laymen it is almost impossible to understand how the "technological wonders" function. Furthermore, the use of technological tools often requires thorough introduction and attendance by experts, thereby increasing the dependency not only on technology itself but also on those who have the knowledge to operate it.

Scientific knowledge is obviously linked to economical interests and power; a fact that is widely illustrated by the attention that is paid to patenting regulations. Differences in the availability of this knowledge and the access to it, lead to "technology gaps" between those profiting from technologies and those who do not. These gaps exist within societies but also between nations on global level and were one reason to launch projects such as Wikipedia and several open access initiatives for scientific literature.[4]

Thus far, we have considered the acquisition of scientific knowledge as desirable and benign. However, this may not always be the case as already described in the 19th century by Mary Shelly in her famous novel "Frankenstein": "*Learn from me, if not by my precepts, at least by my example, how dangerous is the acquirement of knowledge and how much happier that man is who believes his native town to be the world, than he who aspires to become greater than his nature will allow.*" (Shelly, 2003 (1818), p. 54) In this short quotation Victor Frankenstein warns captain Walton against the ambition of acquiring scientific knowledge at any price. Frankenstein himself paid dearly for his discoveries by feeling responsible for the murder of some of his best-loved friends and family members by his own creature. It is interesting to see that Frankenstein not only considers the application of knowledge but already its acquirement as dangerous. On the one hand, Frankenstein points to the imprudent ambition of scientists who do not think about the consequences of their work. On the other hand, he seems to imply that certain knowledge should not be acquired because it will inevitably lead to undesirable applications. Thereby, this novel raises two of the most topical problems of scientific knowledge, that of uncertainty about its consequences and that of dual usage.

Also in contemporary assessments of the social impact of science and technology it is discussed that the same high potential for the development of desirable products can also be applied to design undesirable and dangerous tools. Unfortunately, this dual character is inherent to scientific knowledge, which implies dangers if used for malicious purposes.

A discussion of the impact of scientific knowledge would not be complete without looking at situations, in which we lack important knowledge. Uncertainty is a constant threat as it implies unforeseen side effects.

Two examples for such unintentional consequences are the impact of the fluoro hydrocarbons (CFC) used in commercial and household applications on the ozone layer and global warming caused by the anthropogenic emission of greenhouse gases. In order to avoid such negative surprises, novel technologies are being accompanied by extensive technology assessment procedures. However, even expensive risk assessment cannot predict and thus prevent all negative side effects. Therefore, the risk of negative consequences

---

4 Such a gap has for example been described for digital technology by the OECD (http://www.oecd.org/dataoecd/38/57/1888451.pdf).

always remains, while on the other hand, some predicted dangers may never occur or be much milder than anticipated. Predictions are therefore always in danger to contain type I or type II errors.[5]

How should we deal with these uncertainties in decision-making? Should we be guided by pessimistic scenarios? How does probability of such predictions come into play? The precautionary principle, a widely discussed option, deals with uncertainties according to pessimistic models. Therefore, the lack of certainty for the occurrence of serious damage cannot be used as an excuse to postpone measures to prevent these negative consequences. This concept follows the assumption that type I errors are more dangerous than type II errors, in cases when dramatic consequences are predicted. The Rio Declaration or the law of the European Union are two good examples for this principle. However, the precautionary principle remains controversial indicating, that the dealing with scientific uncertainties remains a highly topical and unsolved problem (Andorno, 2004).

Knowledge as a responsibility of scientists

We assume that the different aspects related to scientific knowledge as discussed above lead to responsibilities of the researchers. However, this statement seems to be controversial, for example the biologist Lewis Wolpert would not agree. In his Wilkins-Bernal-Medawar Lecture 1998 Wolpert claimed that "[…] *reliable scientific knowledge is value-free and has no moral or ethical value. Scientists are not responsible for the technological applications of science; the very nature of science is that it is not possible to predict what will be discovered or how these discoveries could be applied.*" (Wolpert, 2005). However, the majority of authors writing about responsibility of scientist do not share Wolpert's opinion. Hans Jonas as an early example, decidedly advances the view that scientists have a strong responsibility for their findings. According to him, science and its application in technology cannot be separated because: First, science depends on the intellectual feedback from technology. Second, science receives its mandates from technology and finally, science itself depends on technological tools (Jonas, 1987). The philosopher John Forge shares the claim that science and technology are not separable. He argues that all scientists need to realize that science affects people and need to consider potential consequences of their work. Just as someone who did not pay his tax will not be excused when she claims that she did not know that there were tax laws, a scientist cannot ignore that science affects people (Forge, 2000). Therefore, he argues that scientists can also be held responsible for an outcome of their work that they did not foresee, at least in cases in which they did not invest the necessary effort to anticipate potential consequences. It seems as if even Wolpert would agree that scientists are obliged to think about such consequences. Even though scientists are not responsible for the outcome of their work in his opinion, he claims a social obligation of making the implications of their work public (Wolpert, 2005).

5 Type I errors (also called alpha error) reject a correct hypothesis as wrong – false positive. Type II errors (beta error) accept a wrong hypothesis as correct – false negative.

Responsible behavior by scientists

In the light of this debate it is important to discuss how scientists have dealt with anticipated dangers in the past. The importance of such foresight became particularly clear in the 20th century when basic scientific knowledge was directly applied for the construction of the atomic bomb. The development of increasingly efficient nuclear weapons, seriously worried experts in this field, because the use of these weapons was considered to be powerful enough to erase mankind. Thus, in 1955 eleven leading scientists signed the Russell-Einstein Manifesto. In this document they explained in very clear words the disastrous consequences of using nuclear weapons and urged for peaceful means to settle conflicts. Two years later, this manifesto was followed by the first International Pugwash Conference. In this conference, which has been held annually since then, scientists met to appraise the dangers and discuss a resolution urging governments to find peaceful means for the settlement of disputes[6]. The nuclear physicists Joseph Rotblat, one of the signatories of the Russell-Einstein Manifesto, participated for one year in the Manhattan project, which he left on the grounds of conscious. Rotblat was convinced that scientists have a moral responsibility for the application of the knowledge they acquire, culminating in the removal of dangers that might emerge from them: "*I am increasingly concerned about the role of science and technology, both in day-to-day life and in the destiny of humankind. Whether directly through the development of new military capabilities, or indirectly through the uneven distribution of the benefits of new technologies, the future of civilization and the very existence of the human species are imperiled. Scientists bear much responsibility for this danger and must take steps towards its removal. Ethical considerations must become part of the scientist's ethos*" (Rotblat, 2000).

A second important movement of scientists that expressed concerns towards novel technological and scientific developments stems from the life sciences in the context of the development of recombinant DNA technology. Between 1968 and 1973 most of the key tools of this potent technology were established within a short time. Restriction enzymes were discovered, which led to the first recombinant DNA molecule and the first transgenic *Escherichia coli* bacterium. On a Gordon Conference on nucleic acids in 1973 scientists expressed concerns about possible hazards of recombinant DNA. As a result of this conference two open letters by scientists were published in which they stress the unpredictable nature of recombinant DNA technology (Berg *et al.*, 1974; Singer and Soll, 1973). Berg and others point to the risk that novel oncoviruses could emerge and that transgenes, including genes providing antibiotic resistance, might be taken up by natural bacteria. Both letters call for proper evaluation of the risks and ask for clear guidelines on the work with recombinant DNA molecules. Berg *et al.* even call for a voluntary moratorium on certain recombinant DNA experiments until evaluation of potential hazards has been done (Berg *et al.*, 1974; Weiner, 2001). The discussion led to the well-known Asilomar conference in 1975 at which 140 mainly biologists, gathered to discuss the safety of recombinant DNA research. As a consequence of the conference the moratorium was lifted and the NIH (National institutes of health) issued guidelines for institutions that received

---

6 For information about the beginning and the current status of Pugwash conferences see: http://www.pugwash.org/

federal funding for scientific research using recombinant DNA techniques in the US. Following Asilomar the worries among scientists about potentially hazardous consequences of molecular biology quickly declined. Likewise, the NIH guidelines were deregulated soon thereafter. While some authors praised the Asilomar conference as an example for responsible behavior by scientists (Frazier, 1975), others criticized that the conference did not consider ethical and legal implications of genetic engineering or that the main motivation for scientists to discuss these issues was not responsibility but the attempt to avoid governmental regulation (Weiner, 2001; Wright, 2006).

In both, the initiation of the Pugwash and the Asilomar conferences scientists took the initiative to point to risks and hazards related to their discoveries. However, the objectives of the two movements were significantly different. Whereas the Pugwash conference tried to respond to the potential use of nuclear weapons, the Asilomar conference was held at the emergence of a novel technology, whose potential and hazards were unknown. While the first discussed the application of scientific results, the second discussed consequences emerging from the proceeding of science itself. Importantly for our context, both can be seen as examples for responsible behavior of scientists.

However, scientists are not only morally responsible towards society but also towards the scientific community. Hans Lenk phrases this an internal responsibility in contrast to the external responsibility towards the public (Lenk, 1991). This follows from the fact that scientific knowledge is a valuable good for researchers. Science can only function if scientists deal in a responsible way with their own results as well as with findings from others. This implies that they communicate their findings truthfully, acknowledge the participation of colleagues and refer to work of other groups.[7] Other aspects of this responsibility include responsible conduct concerning confident data and assessment of results in peer review. The sociologist Robert Merton had compiled such rules of conduct already in 1942 into what he called "the ethos of science" (Merton, 1973). Today similar points can be found in "guidelines of research integrity" of universities[8], [9] indicating that particularly in the highly competitive environment and under the economical pressure that accompany science today it seems to be important to remind scientist of behaving responsibly towards the scientific community.

After discussing past debates, we will shift our focus now to illustrate and discuss the responsibilities of scientists using a current example: synthetic biology.

Synthetic biology

Before we discuss responsibilities related to synthetic biology we need to give a short introduction to this field and explain why we think that it is a good example for our purpose.

Synthetic biology is an emergent field in which scientific and technological aims and techniques conflate. It includes several different approaches with the common aim

7 The permanent debate about publishing is a reflection of this. A recent example is the editorial (2009), Nat Cell Biol 11, p. 1045.

8 e.g. ETH: http://www.rechtssammlung.ethz.ch/pdf/414_Integrität-Forschung_engl.pdf, university of Oxford: http://www.admin.ox.ac.uk/rso/integrity/

9 We thank Lutz Wingert for the suggestion to compare Merton with ethical guidelines of universities.

to produce new forms of life. Importantly, these novel forms should not be attained by small modifications of existing organisms but rather by a novel overall design. Synthetic biologists use different approaches depending of their scientific background (Deplazes, 2009). The most prevalent approach is the bioengineering subfield, which expands genetic engineering by applying more systematic and extensive genetic modifications in extant organisms introducing rather complete biochemical pathways than single genes (Endy, 2005; Heinemann and Panke, 2006). Another well-known approach is synthetic genomics driven by the Craig Venter institute, which aims to synthesize a genome *in vitro* and replace the extant genome of an organism by the synthetic one. The organism will thus be controlled by the novel genome, which eventually should carry transgenes of useful functions (Holt, 2008).[10] All the approaches are still in their early days and it is not yet clear to what extent the ambitious goals will be attained. Even though synthetic genomics progresses fast (Lartigue, *et al.*, 2009), the approach closest to technological application is bioengineering. However, scientists still struggle with many basic problems such as cellular regulation mechanisms, which reach a complexity that can only be addressed by using computer science with its own separate aims and methodology which might even justify to speak of a separate "*in silico*" approach of synthetic biology (Deplazes, 2009).

Why should we use synthetic biology as an example to illustrate moral responsibility of scientists? Synthetic biology is a suitable candidate for a number of reasons: First, it deals with the type of knowledge that Frankenstein called dangerous, namely the knowledge to design and produce new forms of life.[11] Further, large parts of the public have strong and intuitive feelings that manipulating living organisms asks for more responsibility than the dealing with inert machines; not only because an organism might develop in unintended ways but also, because it somehow touches the essence of our lives as well. More importantly, science and technology are tightly linked in synthetic biology. Jonas' three forms of mutual dependence between science and technology as mentioned above apply perfectly to synthetic biology. First, synthetic biology can only develop further by application of its results. Second, the main driving force for this research is the hope for future applications in technology. Third, the improvement of DNA synthesis technology is one of the factors promoting the progress in synthetic biology. Also the planed applications of synthetic biology provide a good reason to discuss its moral responsibilities. Currently, potential applications comprise the development of new medications or the use of synthetic organisms as a source for renewable energy (Savage *et al.*, 2008). These applications would contain the organisms in defined conditions. However, using synthetic organisms for bioremediation or as biosensors would in most cases include their release into the environment (de Lorenzo, 2008). Furthermore, biosecurity experts are concerned that synthetic biology could be abused for the production of very efficient biological weapons.

Finally, application of synthetic biology comprises a potential to introduce significant changes in human life leading to a huge economic interest in synthetic biology (Schmidt,

---

10 For the discussion of other approaches see e.g. (Deplazes, 2009; Forster and Church, 2007; O'Malley, *et al.*, 2008).

11 Importantly, we do not intend to provoke fears and aversion by referring to the frequently abused "Frankenstein factor". Indeed, we do not imply that synthetic biology is unethical because of its aims. We just like to refer to "Frankenstein" as a book which discusses the related responsibilities.

2008). Altogether, we believe that synthetic biology is in spite of its "young age" a good example to analyze how scientists and other stakeholders deal with responsibilities, also because initial measures towards this direction have already been taken.

Responsibilities for knowledge in synthetic biology

In the following section we will resume some of the topics addressed in the previous sections to illustrate different facets of moral responsibility using the example of synthetic biology and to contribute to a better understanding of the responsibilities linked to this emerging field. We restrict ourselves to the practitioners of synthetic biology as responsibility subjects. However, this does not mean that they are the only subjects holding responsibility. We will discuss responsibility concerning three different functions of knowledge in synthetic biology: First, knowledge as a source of synthetic biology, second, knowledge as a product of synthetic biology and finally, knowledge about consequences of synthetic biology.

- As discussed, knowledge gained by others, is an important source for scientific findings. This is true for synthetic biology as it is for other sciences. However, the access to knowledge seems to be at stake by the conflicting interests of economy, according to which important findings should be patented. This conflict is even more enforced by a general tendency to patent also basic knowledge. An example from synthetic biology is the application for the patent on a minimal bacterial genome by the Craig Venter institute in 2007 (Kaiser, 2007). The patent claims rights on basic ideas and techniques in synthetic biology, which has caused opponents to accuse the Venter Institute of trying to create a monopoly. However other voices claim that patents warrant the access to knowledge (once license fees are paid) whereas alternatively, results would be concealed (Schmidt *et al.*, 2008). In particular due to the conflicting economical interests which are reflected in this controversy, it is important to keep the internal responsibility in synthetic biology in mind.

- Knowledge is the first product of science. For synthetic biology this is the knowledge about the design and production of living organisms. So far most techniques used in synthetic biology start from existing organisms, and products of synthetic biology may not be considered "new form of life", yet. However, already the claim that this goal is within reach and even more its potential achievement in the future will have a high impact on our understanding of life and of nature (Deplazes and Huppenbauer, 2009). Some religious positions may argue that humans should not see life as a product and that this knowledge" should therefore not be strived for, a similar attitude is also expressed by "Frankenstein". However, this raises the question whether responsibility of scientists towards society implies the needs to justify or explain why they want to acquire this type of knowledge. Normative criteria such as for example religious rules do not offer a solution, since many scientists do not consider those to be binding. However, if a scientist does not behave responsibly according to one set of rules this does not mean that she is acting irresponsibly based on some other rules. It will thus be important to decide which rules are binding for scientists and evaluate which responsibilities are based on only these rules. According to the "dual-use" dilemma knowledge can be considered dangerous if it can be applied to cause harm. This is certainly the case for knowledge of synthetic biology techniques, which implies the potential design of biological weapons. The example of Pugwash as discussed above has shown how responsible scientist can inform about the dangers of their findings. A comparable approach has in fact already been taken by synthetic biologists as will be discussed at the end of this section.

- We have already discussed that responsibility can also arise from lack of knowledge as in case of uncertainties and unforeseen consequences. Biotechnologies are particularly prone to this problem because they handle living organisms, which reproduce and can act unexpectedly. Furthermore, through selective advantages of certain mutants new and unexpected features, such as resistances, can evolve. It is thus important to consider safety measures against potential side effects of synthetic biology. However, the assessment of side effects of synthetic biology is complicated by the multidisciplinarity of the field. Engineers and computer scientists, who design complex pathways, might not be aware of potential biological consequences and for biologists, unfamiliar with engineering strategies, it might be difficult to predict the complexity of the engineered metabolism and its consequences. Therefore, interdisciplinary collaborations and multidisciplinary teams as well as educational programs combining different skills seem to be promising approaches for a successful and farsighted implementation of synthetic biology and its assessment.

Table 1 The different elements of responsibility in three functions of knowledge in synthetic biology SB stands for synthetic biology.

	Subject	Object	Instance	Normative criterion
Knowledge as a source of SB	Synthetic biologists	Basic knowledge	Scientific community	Guidelines of research integrity
Knowledge as a product of SB	Synthetic biologists	"Problematic" or "dangerous" knowledge	Society, nature, religion	Social rules, Ethical theory, Religious rules
Knowledge about consequences of SB	Synthetic biologists	Consequences of SB	Society, environment	Precautionary principle Guidance by most likely scenarios

In table 1 we summarize the different elements of responsibility related to the three functions of knowledge in relation to the subject synthetic biologist. The table indicates that several issues need to be clarified before responsibilities can be assigned. It is relatively easy to determine the object of responsibility, since this aspect is generally the most discussed one. However, the instance towards which a synthetic biologist is responsible and its underlying normative criterion are more critical.

Synthetic biology with its aims and ambitions is a young "discipline" of the 21st century. Thus, it is interesting to observe how this scientific community handles the responsibilities arising from their work after experiences such as Asilomar and the GMO debate in Europe. In the preparation of SB 2.0, the second in a series of international synthetic biology meetings, a draft of a document describing a self-governance process for synthetic biology has been sketched. The initial plan was that, in analogy to the Asilomar conference, all member of the SB 2.0 conference would discuss, adapt and sign the document[12]. However, the idea of self-governance for synthetic biology raised objection and a group of 38 international organizations signed an open letter in which they called *"on synthetic biologists to abandon their proposals for self-governance and to engage in an inclusive process of global societal debate on the implications of their work"*.[13] The authors of this letter asked for "*strong mandatory measures in accordance with the precautionary principle*" and they demanded that society should be engaged in a dialogue. Furthermore, they stressed that many synthetic biologists have a conflict of interest because of shares in companies in the field of synthetic biology. Subsequently, the draft written for the SB 2.0 conference has never been submitted for signature. However, societal and ethical impacts of synthetic biology have ever since been part of discussions about synthetic biology. Several private and governmental projects address these issues. They are part of the major synthetic biology conferences and are discussed in scientific journals. Ethicist and social scientists apparently play an important role for the further trends of synthetic biology. At the last conference of the SB series, SB 4.0, non-governmental organizations such as the ETC (Action Group on Erosion, Technology and Concentration), which is known for its critical attitude towards synthetic biology, was given the opportunity to present their views. Thus, the synthetic biology community exemplarily supports a broadly based assessment of synthetic biology at a very early stage of the technology and provides a platform for a dialogue between different stakeholders. However, at the SB 4.0 conference where different panels took place at the same time, it was noticeable that in the panels on societal and ethical issues social scientists and ethicists were generally almost among themselves whereas most synthetic biology practitioners attended other panels. This raises the question, whether it is possible to delegate moral responsibility. Is it enough for synthetic biologists to involve social scientists and ethicists? Can responsibility be passed on to the "responsibility-specialists"? Such a solution would certainly be counter-productive. Synthetic biology is a typical example for a field that raises collective responsibilities, which are held by different stakeholders and cannot be transferred. Scientists and engineers from different countries, different institutions and disciplines as well as social scientists, ethicists and non-governmental organizations have their own part of moral responsibility. It will be interesting to observe whether the first impression is true and the multi stakeholder approach indeed influences the further development of synthetic biology and its applications and if so, what the effects of this influence will be.

Conclusion

As outlined so far, scientific knowledge can have ethical impact. Scientific knowledge can be a right if we think of it as the source of research on which researchers depend. However, scientific knowledge can also be a duty when scientists need to think about potential consequences of their work as suggested by Forge (Forge, 2000). We have discussed that such knowledge can be dangerous if it can be abused for malicious purposes. The different roles

12 The draft by S. Maurer, K.V. Lucas and S. Terrell has been published on line as appendix A of: http://gspp.berkeley.edu/iths/UC%20White%20Paper.pdf

13 www.etcgroup.org/upload/publication/pdf_file/8

of scientific knowledge are related to the moral responsibilities of synthetic biologists, which differ concerning the *object*, the *instance* or the *normative criterion*. We have discussed how scientists have taken action and addressed societal aspects of their knowledge. Even though several aspects of the Asilomar conference or the self-governance approach of synthetic biologists have legitimately been criticized, they do provide evidence that some scientists are conscious the impact of their research. Some of them may try to avoid public reactions or governmental regulations, but interviews with other scientists indicate that they are or were genuinely concerned.

Due to its promises synthetic biology is currently associated with high expectations. Although there has not been a strong public reaction against it, yet, scientists have realized that synthetic biology has the potential to be extremely contentious. The open letter signed by international organizations as a reaction to the suggested self-governance is one of the first indications for such a tendency. Two social scientists observing the development of synthetic biology: Jane Calvert and Paul Martin wrote that one response to concerns as those expressed in the open letter was "*to institutionalize the involvement of social scientists in the field*" (Calvert and Martin, 2009). Does this mean that also the related responsibilities can be institutionalized? We do not think that it is possible to delegate responsibility to social scientists and ethicists. They can analyze synthetic biology from an outside perspective; they can give advice what responsible behavior might be and thereby take a share of responsibility in the further development of synthetic biology. However, in the end each and every scientist is responsible for her dealing with scientific knowledge and for the decision which risks or compromises he thinks are morally acceptable.

References

Editorial (2009) "Accurately reporting research", *Nat Cell Biol* 11, p. 1045.

Andorno, R. (2004) "The Precautionary Principle: A New Legal Standard for a Technological Age", *Journal of International Biotechnology Law* 1, p. 11-19.

Berg, P. *et al.* (1974) "Potential biohazards of recombinant DNA molecules", *Science* 185, p. 303.

Calvert, J. and Martin, P. (2009) "The role of social scientists in synthetic biology. Science & Society Series on Convergence Research", *EMBO Rep* 10, p. 201-204.

De Lorenzo, V. (2008) "Systems biology approaches to bioremediation", *Curr Opin Biotechnol* 19, p. 579-589.

Deplazes, A. (2009) "Piecing together a puzzle. An exposition of synthetic biology", *EMBO Rep* 10, p. 428-432.

Deplazes, A. and Huppenbauer, M. (2009) "Synthetic Organisms – Living Machines, Positioning the products of synthetic biology at the borderline between living and non-living matter", *Syst Synth Biol.*

Endy, D. (2005) "Foundations for engineering biology", *Nature* 438, p. 449-453.

Eshleman, A. (2008) "Moral Responsibility", in *The Stanford Encyclopedia of Philosophy*, Zalta, E. N. (Ed), <http://plato.stanford.edu/archives/fall2008/entries/moral-responsibility/>.

Forge, J. (2000) "Moral responsibility and the 'ignorant scientist'", *Sci Eng Ethics* 6, p. 341-349.

Forster, A. C. and Church, G. M. (2007) "Synthetic biology projects in vitro", *Genome Res* 17, p. 1-6.

Frazier, K. (1975) "Rise to Responsibility at Asilomar", *Science News* 107, p. 187.

Heinemann, M. and Panke, S. (2006) "Synthetic biology-putting engineering into biology", *Bioinformatics* 22, p. 2790-2799.

Holt, R. A. (2008) "Synthetic genomes brought closer to life", *Nat Biotechnol* 26, p. 296-297.

Jonas, H. (1987) "Warum die Technik ein Gegenstand für die Ethik ist: Fünf Gründe", p. 81-92, in *Technik und Ethik,* Lenk, H. and Ropohl, G. (Eds), Stuttgart, Reclam.

Kaiser, J. (2007) "Synthetic biology. Attempt to patent artificial organism draws a protest", *Science* 316, p. 1557.

Lartigue, C., Vashee, S., Algire, M. A., Chuang, R. Y., Benders, G. A., Ma, L., Noskov, V. N., Denisova, E. A., Gibson, D. G., Assad-Garcia, N., Alperovich, N., Thomas, D. W., Merryman, C., Hutchison, C. A.,

3RD, SMITH, H. O., VENTER, J. C. and GLASS, J. I. (2009) "Creating bacterial strains from genomes that have been cloned and engineered in yeast", *Science* 325, p. 1693-1696.

LENK, H. (1991) "Zu einer praxisnahen Ethik der Verantwortung in den Wissenschaften", p. 54-76, in *Wissenschaft und Ethik,* LENK, H. (Eds), Stuttgart, Reclam.

LENK, H. and MARING, M. (2001) "Problems of responsibility in and for technology", p. 441-458, in *Advances and Problems in the Philosophy of Technology,* LENK, H. and MARING, M. (Eds), Münster, Hamburg, London, Lit Verlag.

MERTON, R. K. (1973) *The Sociology of Science. Theoretical ane Empirical Investigations*, Chicago/London, The University of Chicago Press.

O'MALLEY, M. A., POWELL, A., DAVIES, J. F. and CALVERT, J. (2008) "Knowledge-making distinctions in synthetic biology", *Bioessays* 30, p. 57-65.

ROPOHL, G. (1987) "Neue Wege, die Technik zu verantworten", p. 149-177, in *Technik und Ethik,* LENK, H. and ROPOHL, G. (Eds), Stuttgart, Reclam.

ROTBLAT, J. (2000) "ESSAYS ON SCIENCE AND SOCIETY: Taking Responsibility", *Science* 289, p. 729.

SAVAGE, D. F., WAY, J. and SILVER, P. A. (2008) "Defossiling fuel: how synthetic biology can transform biofuel production", *ACS Chem Biol* 3, p. 13-16.

SCHMIDT, M. (2008) "Diffusion of synthetic biology: a challenge to biosafety", *Syst Synth Biol* 2, p. 1-6.

SCHMIDT, M., TORGERSEN, H., GANGULI-MITRA, A., KELLE, A., DEPLAZES, A. and BILLER-ANDORNO, N. (2008) "SYNBIOSAFE e-conference: online community discussion on the societal aspects of synthetic biology", *Syst Synth Biol* 2, p. 7-17.

SHELLY, M. (2003 (1818)) *Frankenstein*, London, Penguin Books.

SINGER, M. and SOLL, D. (1973) "Guidelines for DNA hybrid molecules", *Science* 181, p. 1114.

SMILEY, M. (2008) "Collective Responsibility", in *The Stanford Encyclopedia of Philosophy,* Zalta, E. N. (Ed), <http://plato.stanford.edu/archives/fall2008/entries/moral-responsibility/>.

WEINER, C. (2001) "Drawing the line in genetic engineering. Self-regulation and public participation", *Perspect Biol Med* 44, p. 208-220.

WOLPERT, L. (2005) "The Medawar Lecture 1998: Is Science Dangerous?", *Philosophical Transactions: Biological Sciences* 360, p. 1253-1258.

WRIGHT, S. (2006) "DNA Technology: Asilomar Conference and 'Moratorium' on Use", in *Encyclopedia of Life Sciences,* (Ed), John Wiley & Sons, Ltd. www.els.net.

ZIMMERLI, W. C. (1987) "Wandelt sich die Verantwortung mit dem technischen Wandel?", p. 92-112, in *Technik und Ethik,* LENK, H. and ROPOHL, G. (Eds), Stuttgart, Reclam.

Chapter 2

Common Goals

2.1 War and Peace: Conflict and Cooperation in a Tropical Insect Society

by Raghavendra Gadagkar[1]

It is lamentable, that to be a good patriot one must become the enemy of the rest of mankind.

Voltaire (1694-1778), Philosophical Dictionary

The Insect Societies

Most insect species appear to complete their life cycles without the need for any significant interaction with other members of their species except during the act of mating. Other species however, practice group life with varying degrees of dependence on interaction with conspecifics[2]. The most extreme of such dependence is seen in the so called eusocial species that spend all or most of their lives in colonies. These include some bees and wasps and all ants and termites. We often refer to these social insect species as insect societies. Indeed, as we will see below, the insect societies rival if not surpass human societies in the complexity of their social organization and integration, division of labor, communication and even their caste systems. A matter of great interest, but probably of little relevance to our present discussion is that (with the exception of the termites) the insect societies are "feminine monarchies". I borrow the phrase first used by the cleric Charles Butler in 1634 to describe the honeybee society; their colonies consist of queens and female workers while the males play no domestic role – they merely mate and die. The most striking feature of insect societies is reproductive division of labor – only one or a small

1 Raghavendra Gadagkar is INSA SN Bose Research Professor and JC Bose National Fellow at the Centre for Ecological Sciences and is founding Chairman of the Centre for Contemporary Studies, Indian Institute of Science, Bangalore.

2 Other members of their own species.

number of individuals reproduce (the queens) while the rest remain sterile (the workers) and perform all the tasks associated with nest building and maintenance, foraging and brood care. In addition to reproductive division of labor between the queens and the workers, there is often further division of non-reproductive labor among the workers. It is this division of labor, first between the queen caste and worker caste and then between the different worker sub-castes that appear to be the secret of the unparalleled ecological success and dominance of social insects as compared to non-social species, insect or otherwise (Wilson, 1971; Wilson, 1990; Hölldobler and Wilson, 2009).

Although all dichotomies will eventually break down under careful scrutiny, it is useful to recognize two sub-divisions among the eusocial species, the primitively eusocial and the advanced eusocial. The following is a brief and somewhat oversimplified description of our current understanding of the contrast between primitively and highly eusocial species. The primitively eusocial species live in relatively small colonies (<100 adults) with limited division of labor so that neither queens and workers nor the sub-castes among workers are morphologically differentiated. Relative to the highly eusocial species, these colonies are organized through top-down control by a physically aggressive queen. In contrast, the highly eusocial species live in large colonies (sometimes up to a million adults) and are largely self-organized through bottom-up regulation in a decentralized manner through worker-worker interactions. In primitively eusocial species the physically stronger queen suppresses worker reproduction and also drives the workers to labor for the colony through physical aggression and harassment of the workers. In highly asocial species queens produce and release chemicals (pheromones) that are a reliable signal of their superior fertility. The workers are selected to "obey" and suppress their own reproduction as it is in their best fitness interests to do so. It is also in their own best interests to efficiently organize their labor in the colony since their fitness now derives from efficient reproduction of their mother queen rather than from their own reproduction (Wilson, 1971; Hölldober and Wilson, 2009; Keller and Nonacs, 1993), (Figures 1-3).

A Tropical Primitively Eusocial Wasp Society

There occurs quite commonly in urban areas of southern India, a remarkable primitively eusocial wasp. Known scientifically as *Ropalidia marginata* (Figure 4), we refer to this commonly as the Indian paper wasp. The name paper wasp comes from the fact that these (and nearly all social wasps) construct their nests, not from wax as bees do, nor from leaves as many ants do or from soil as termites do, but from paper – paper which they manufacture by scraping cellulose fibers from plants, adding their own secretions and chewing it into a fine pulp. Apart from being quite common, there are many features of this wasp that make it an excellent model system for our research. Unlike many other social wasps, they build small open nests so that all adults and brood, and more importantly, all social interactions, are easily visible to the human observer. The number of adult wasps in a nest is often in the 20's and 30's and almost never exceeds a 100. The small number of wasps and their easy visibility make it possible to individually mark[3] all the wasps in a colony and

3 With small spots of quick drying, non-toxic, coloured paint of different colours, on different parts of their body.

Figure 1 The honeybee is the most widely known example of a highly eusocial insect society. Shown here is the nest of perhaps the least advanced of the honeybees, the Asian dwarf honeybee, Apis florea. (Photograph: Dr. Thresiamma Varghese). The underlying wax comb, brood and food stores are all covered by the large numbers of bees on the nest surface.

Figure 2 A close up of a portion of the honeybee nest on figure 1, showing the morphologically differentiated queen surrounded by a group of workers. (Photograph: Dr. Thresiamma Varghese).

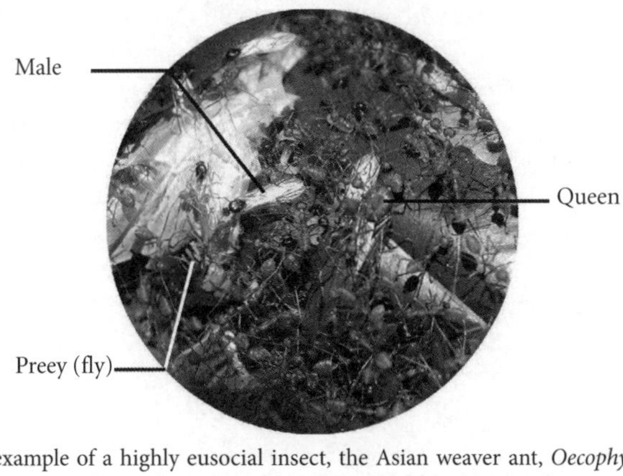

Figure 3 Another example of a highly eusocial insect, the Asian weaver ant, *Oecophylla smaragdina*, showing the large queen, some males and an item of prey (a fly) brought into the nest as food. (Photograph: Dr. Thresiamma Varghese).

observe them throughout their lives. The nest is a small honeycomb like structure with typical hexagonal cells that are used by the wasps to rear their brood.

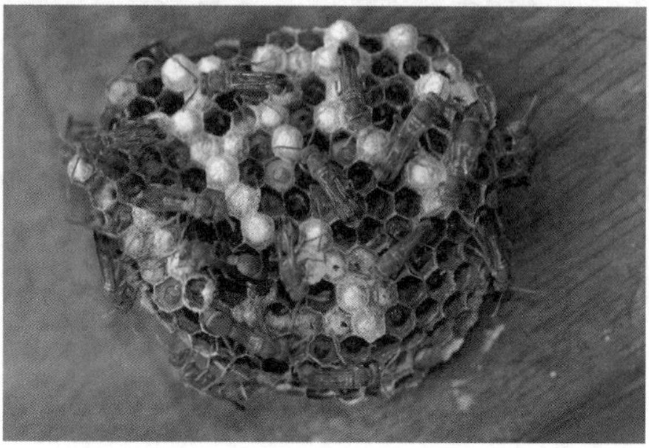

Figure 4 A typical nest of the primitively eusocial wasp, *Ropalidia marginata*, showing the paper comb with hexagonal cells, many adult wasps and some capped brood (pupae). (Photograph: Dr. Thresiamma Varghese).

New nests may be founded either by a single female wasp or by small groups of them. In the single-foundress nests the lone female naturally does everything by herself – she builds the nest, lays eggs, forages for food and feeds her larvae when they hatch, guards them when they pupate and rears them to adulthood unaided, thus combining the duties of queen and worker. In multiple-foundress nests there is a clear division of labor. One individual lays eggs while the others work toward nest building, foraging and brood care. We designate the egg-layer as the queen and the remaining wasps as workers, even though (to us) they appear morphologically indistinguishable from each other. The first adult wasps to eclose[4], both in single as well as in multiple-foundress nests are usually female wasps and normally stay back in their nest of birth and function as workers. Males are produced later and they stay on in their nest of birth only for about a week. The week-old males depart to lead a nomadic life, attempting to mate with female wasps from various other nests that may be out in search of food or cellulose fibers for building their nests. Thus our systematic observations are generally concerned only with female wasps on the nest.

Female wasps have at least four different life-style options available to them. They may leave their nest of birth and start single-foundress nests of their own, they may leave and join other multiple foundress nests, they may stay back in the nest of their birth and spend their entire lives functioning as sterile workers and finally, they sometimes work for a while and then, at an opportune moment, replace the original queen as the new queen of the colony. Workers appear to mate opportunistically during their foraging flights so that some but not all workers in a colony are mated. Mated females store sperm received from one or more males, in their spermatheca[5] and can use them when required. Female

4 Emerge as adults from the pupa, after metamorphosis.

5 A small pouch that is connected to the oviduct and has several accessory glands associated with it so that the sperm are nourished and kept viable for the duration of the life of the individual, which incidentally can be a few decades in some ants.

wasps can lay both unfertilized, haploid[6] eggs as well as fertilized diploid[7] eggs. The unfertilized eggs develop parthenogenetically into adult males while the fertilized eggs develop into adult females. Thus mothers have complete control over the sex of their offspring, producing sons and daughters as required. The first brood in single as well as in multiple-foundress colonies is generally female and the ensuing daughters generally stay back as workers and assist their mother in producing more brood. Workers perform all domestic tasks in an orderly fashion, working intranidally (at home) at brood care, at nest building and maintenance when relatively young and away from home, and at gathering food and cellulose fibers when relatively old. Workers live on an average for about four weeks although their life spans range from a few days to several months. Queen tenures are also highly variable, ranging from just about a week to nearly a year.

I have spent the past 30 years studying these wasps, often in collaboration with a large number of enthusiastic students. My research has been driven entirely by my insatiable curiosity about how these wasps organize their lives, coordinate their activities and manage to function as an orderly and efficient society. In other words, I wish to understand the rules that govern the lives of these wasps and the mechanisms they employ to create, follow and perhaps enforce these rules. As an evolutionary biologist I am simultaneously interested in the evolutionary forces that mould the social and altruistic behavior that these wasps display. This is an especially fascinating conundrum because most of the wasps that impress me with their social accomplishments are sterile and therefore not subject to simple natural selection through "survival of the fittest". We have made many discoveries and learnt a great many odd and amazing facts about these wasps (Gadagkar, 2001). It is hard to pick a winner but perhaps the most fascinating aspect of the lives of these wasps is the fine balance that they tread between conflict and cooperation, and the manner in which they are able to use a subtle mix of conflict and cooperation to achieve success as a society. This is therefore the aspect that I will focus on for the rest of this chapter. One of our most interesting findings in this context is that these wasps display extreme aggression and intolerance toward conspecifics that do not belong to their colonies. However, the wasps are highly tolerant of each other and display almost no aggression to members of their own colony, even when there is a great potential for conflict. Here I will describe and contrast such "war" with foreigners and "peace" with insiders displayed by the wasps. I will also illustrate our research methodology that permits an understanding of these insect societies and conclude with some remarks about what we might learn from these primitive insect societies.

Nestmate Discrimination

Seeing the wasps flying in and out of their colonies[8] without any apparent hesitation, even the casual observer will wonder how the wasps know which is their colony and which is not. This ability becomes even more intriguing when we see that two or more colonies are sometimes built very close to each other. Is it indeed that the wasps know exactly which

6 Containing only one set of [maternal] chromosomes.

7 Containing two sets of chromosomes, one maternal and one paternal.

8 I will use colony and nest somewhat interchangeably.

their nest is and never make mistakes or is that they sometime enter the wrong colony? The casual observer is unlikely to find out. But anyone who observes these wasps longer than the casual observer is likely to be rewarded by occasional breakdown of this apparent orderly behavior. Wasps do occasionally land on alien nests, either because they have made a mistake or because they wish to deliberately sneak into a foreign colony. Whatever be the reason for the unusual action, a drama ensues that would rivet anyone's attention. The intruder is intensely and aggressively repelled by one or more resident wasps from the colony on which she has landed. The intruder quickly beats a retreat and if she fails to do so she may be mauled, stung and even killed. How does this happen? How do the wasps recognize and repel the intruder? In other words, how do the wasps discriminate between their nestmates and non nestmates? We decided to investigate these questions using an experimental approach (Gadagkar, 1985).

In the first set of experiments we put three wasps in a plastic jar and recorded all behavioral interactions shown by each of them toward each of the other two. We called this the triplet assay. The three wasps were so chosen that two were nestmates of each other and one was a non nestmate of the other two. We observed 15 types of interactions among the wasps including antennation, approach, nibble, peck, chase, attack and so on (for a list of the 15 behaviors and their tolerance ranks[9], see Table 1). Considering how tolerant or intolerant these behaviors seemed to be and how often each type of behavior was shown, we computed a numerical index of tolerance. The tolerance index was computed by multiplying the proportional representation of each individual behavior (such as nibble, peck etc.) multiplied by the tolerance rank of that behavior and then summing over all the behaviors shown. Since each of the three wasps can initiate interactions with each of the other two, we could compute six tolerance indices per experiment (Figure 5) and then compute separate average tolerance indices for all interactions among nestmates and all interactions among non nestmates. Thus we could make a quantitative assessment of whether individual wasps were able to discriminate their nestmates from non nestmates, based on statistical significance of differences in the tolerance indices among nestmates and those among non nestmates.

In the first experiment, we collected nests from nature, brought them to the laboratory and transferred the adults already present at the time of collection into individual plastic jars and kept them in isolation for several days (See Figure 6 for an illustration of the plastic jars and wood and wire mesh cages used in these experiments). We collected nests from some 10 km apart from each other so that we had both nestmate wasps and non nestmate wasps for the experiment. We then tested these wasps by putting two nestmates and one non nestmate into a fresh plastic jar, allowed them to interact, constructed tolerance indices and compared the tolerance shown toward nestmates and non nestmates. There was clear evidence of nestmate discrimination, i.e., nestmates were significantly more tolerant of each other than they were of their non nestmates. These wasps were able to successfully recognize their nestmates either because they were genetically related, or because they had shared the same nest environment or simply because they were directly familiar with nestmates they encountered in the experiment.

[9] Our tolerance ranking was based on more than 200 hours of observation of wasps on colonies in nature and in the laboratory.

Table 1 The 15 behaviors observed in the Triplet assay, ranked by increasing order of tolerance. Reproduced with permission (Venkataraman, *et al.*, 1988).

Tolerance Rank	Behavior
1	Aggressive bite: the most extreme form of aggression seen in this species; sometimes leads to injuries
2	Attack: a ritual act of aggression where the dominant animal climbs onto the subordinate and attempts to bite its mouthparts
3	Peck
4	Chase
5	*[10]Aggressive mutual antennation: a kind of sparring contest
6	Nibble: relatively mild with little chance of injury
7	Crash: the crashing of one wasp into a sitting wasp that results in one or both falling to the ground; very brief and appears much milder than behaviors 1 - 6
8	*Falling fight: two animals grappling with each other and falling to the ground; very brief and appears much milder than behaviors 1 - 6
9	Avoiding
10	Soliciting
11	*Mutual approach with withdrawal
12	Approach I: the other withdraws
13	Approach II: the other does not withdraw
14	Antennation
15	*Mutual antennation

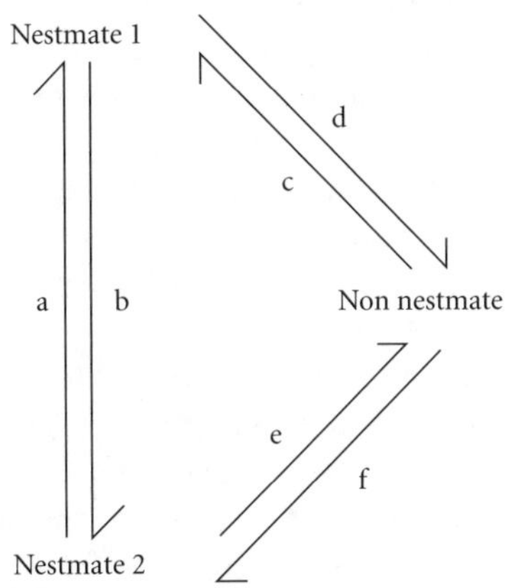

Figure 5 The triplet assay. The behavioral interactions seen in all experiments are classified into six categories, designated a, b, c, d, e, f, such that all interactions initiated by nestmate 2 towards nestmate 1 are assigned to a, all interactions initiated by nestmate 1 towards nestmate 2 are assigned to b, and so on. Tolerance indices were computed for each category of interaction as described in the text. Reprinted with permission (Venkataraman *et al.*, 1988).

[10] Indicates bidirectional behavior.

To tease out the effects of these factors on nestmate discrimination, we tested wasps that were genetically related and had the experience of sharing the same nest but were not directly familiar with each other. These wasps also showed clear evidence of nestmate discrimination suggesting that direct familiarity was not necessary for discrimination. Thus to be tolerant to each other it is sufficient for both to have had some experience with their nest of birth, even at different times, without coming in direct contact with each other. But is experience with the same nest necessary for two wasps to recognize each other as nestmates? To test this we used wasps that had been removed from their nests just before they completed their development and had been allowed to complete their development in an incubator. These wasps that did not have the opportunity to smell their nests or nestmates or acquire any information about their nests or nestmates during their adult life, completely failed to discriminate nestmates from non nestmates. We thus showed that familiarity with their nest of birth was both necessary and sufficient for discrimination. We labeled the wasps that had experience with their nests and were capable of discriminating their nestmates from non nestmates as "experienced wasps" and those that did not have such experience necessary for making nestmate discrimination as "naïve wasps". Thus experienced wasps were able to tell their nestmates apart from their non nestmates even outside the context of their nests and even if they had not encountered their nestmates before. What do experienced wasps possess that naïve wasps do not, which permit nestmate discrimination? We can imagine that effective nestmate discrimination requires that wasps carry some form of label in their bodies (most likely a smell) which indicates which nest they come from and a template in their head indicating what wasps of their nest should smell like. Experienced wasps obviously possess both labels and templates characteristic of their nests while naïve wasps must lack appropriate labels or templates or both. Whether they lack both labels and templates characteristic of their nests or only one of them is unclear from these experiments and somewhat irrelevant at this point in time, although it will become relevant in the next section. These experiments tell us how wasps are able to efficiently repel non nestmate intruders without interfering with the smooth arrival and departure of nestmates. And that is important because on all days with good weather there is a steady traffic of nestmates going in and out of the nest.

On the few occasions that we have been fortunate to see non nestmates land on nests, we have always witnessed efficient repulsion of their attempts to enter a foreign nest and we have never seen a returning nestmate being subjected to anything of the kind of inspection and harassment meted out to non nestmates. Nevertheless we only very rarely witness landing of non nestmates, in nature. We therefore decided to simulate such landings in the laboratory and in somewhat more natural conditions than the previous set of experiments in the plastic jars. We placed two colonies in two large laboratory cages (See Figure 6) and we introduced all the wasps from one of the cages (without their nest or brood) into the cage of the other nest. Now there were several attempts by the introduced wasps to land on the nest in the cage they were introduced into. The response of the resident wasps to the introduced non nestmate wasps was most interesting. The queen among the non nestmate was selectively attacked, dismembered, and killed within a few minutes of her introduction. In contrast the relatively young ones among the introduced non nestmates were not only permitted to land but were actually accepted and allowed to join the non nestmate colony. The rest of the introduced non nestmate wasps that were neither the queen nor were very young, were allowed to live in the periphery of the cage without being killed and also without being accepted onto the nest (Venkataraman and Gadagkar, 1993). The treatment meted out to the non nestmate queen was spectacular but not really surprising. A non nestmate queen would be dangerous indeed and it would not be worth taking chances with her; she might enter the nest at some moment when the resident wasps lower their vigil and start laying eggs. If she does that then the workers of the resident nest will have to care for the brood of an unrelated individual and this would yield them no evolutionary fitness. That the older workers among the non nestmates were not allowed to land on the nest was also not surprising as it was consistent with what we have observed in nature. That the young non nestmate wasps were accepted onto the alien nests was however rather surprising.

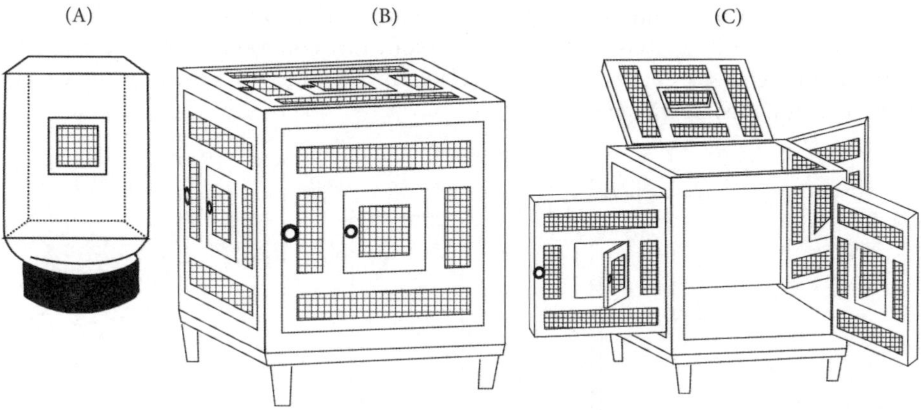

Figure 6 A: Plastic box of dimension 22 × 11 × 11 cm, used to isolate individual wasps and conduct the triplet assay described in the text. B and C: Wood and wire mesh cages of dimension 45 × 45 × 45 cm used for rearing large colonies of *Ropalidia marginata*, and for conducting the experiments in semi-natural conditions described in the text. Reprinted with permission (Gadagkar, 2001).

To understand the factors affecting the acceptance of alien wasps we conducted additional experiments in which we introduced into cages containing nests, many non nestmates drawn from other nests; this time we deliberately chose non nestmates of different ages ranging from 0 to 29 days after their own emergence as adults on their nests (Venkataraman and Gadagkar, 1995). These experiments confirmed that young wasps equal to or less than eight days of age had a high probability of being accepted onto alien nests. This is likely to be related to the fact that younger wasps are expected to have poorly developed labels and templates compared to older wasps. But from an evolutionary point of view, what is the advantage for the resident wasps of accepting young non nestmates and of rejecting older non nestmates? In a new set of experiments we considered two possibilities. One is that young wasps have very poorly developed ovaries and older wasps, even though not the queens of their colonies, sometimes have partially developed ovaries. For the same reason that the non nestmate queens were killed, it may be best not to accept older non nestmates lest they also enter and start laying eggs at some future point in time. The other possibility is that it is age itself that matters and not differential ovarian development associated with age. In experiments designed to dissociate the tangled effects of age and ovarian development, we showed that age is a more important predictor of acceptance than ovarian development (Arathi *et al.*, 1997a). We suspect that younger non nestmates are accepted more often because in addition to having poorly developed ovaries, they may also be more easily molded into working for their foster colonies. Conversely, older individuals may be accepted less often not only because they have partially developed ovaries but also because their behavior may be more difficult to mould in new directions. It must be pointed out that young wasps made more attempts to join the alien nest available to them in the experiment and also met with greater probability of success per attempt in being accepted, suggesting that both the joiners and the acceptors had an interest in their joining (Venkataraman and Gadagkar, 1995) (Figure 7). We do not know how often young non nestmates land on alien nest and provide opportunities for alien colonies to accept and mould them, but the possibility cannot be ruled out. In summary, social insects may occasionally wage war with other species that may be their prey or their predators but their real, day-to-day concern is the threat of colony invasion by other members of their own species. Hence they have developed an elaborate machinery of chemicals and behavior to maintain the integrity of their colonies.

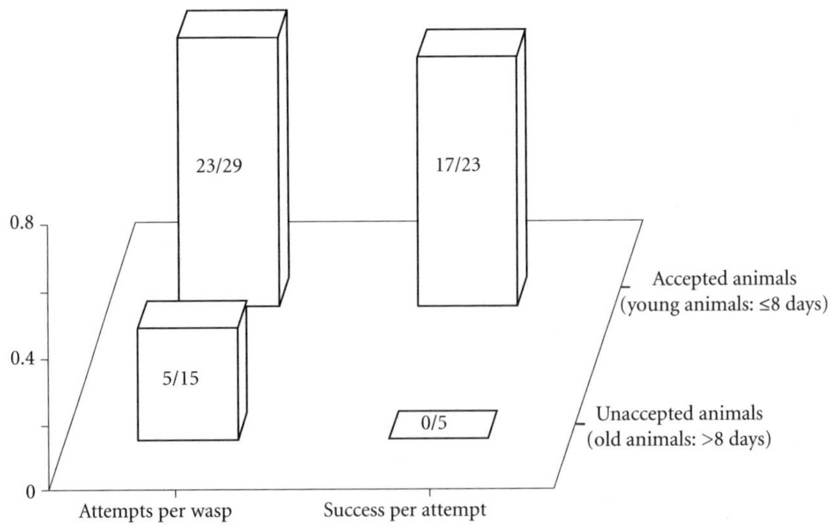

Figure 7 Young introduced wasps attempt to join alien nests and get accepted. The bars on the left represent the number of attempts per wasp, for wasps less than and equal to 8 days and for wasps greater than 8 days. The bars on the right represent the number of successful attempts/total attempts for wasps less than or equal to 8 days and wasps greater than 8 days. Wasps less than or equal to 8 days make significantly more attempts/individual than wasps greater than 8 days (Monte-carlo test, $p < 0.02$) wasps less than or equal to 8 days also have a higher success rate per attempt than wasps greater than 8 days (Monte-Carlo test, $p < 0.001$). Reprinted with permission (Venkataraman and Gadagkar, 1995).

Intra-colony Kin Recognition

As we have seen the wasps are very efficient at detecting and keeping outsiders away, but is there also strife within the colony, among relatively different insiders? Most social insect colonies are family affairs consisting of the queen and her daughters. And yet there is plenty of scope for intra-colony heterogeneity in genetic relatedness. First, workers who are usually the queen's daughters are of course not clones of each other and share only 75% of their genes and should therefore be prone to some conflict. Here the genetic relatedness between two full sisters is not 50% as in diploid organisms but 75% because males are haploid and contribute an identical set of genes to each daughter. The females being diploid, contribute 50% of their genes to their daughters. Thus full-sisters on an average share 75%[11] of their genes with each other. Second, the queen can mate with two or more males and simultaneously produce half-sisters who are related to each other by only 0.25%[12], providing even greater scope for intra-colony conflict. We investigated this phenomenon in *R. marginata* and showed that queens mate with from one to three different males and the average intra-colony relatedness thus drops from the theoretically expected 75% (for single partner mating) to nearly 50% (Muralidharan *et al.*, 1986). This should of course offer more scope for conflict. The value 50% for average intra-colony relatedness actually belies the extent of conflict that is possible because it obscures the fact the colony may contain up to three separate groups of half-sisters related among themselves by 75% but between each other by only 25%. In *R. marginata* we found evidence of even more genetic heterogeneity. Recall that our earliest experiments suggested wasps have the option of working for some time in their nest of birth and then replacing the original queen. It turns out that wasps really do exercise this option from time to time. In a long-term study of four colonies we found that queens maintained their status for variable periods of time, ranging from 7 to 219 days and were replaced at the end of their tenures by their daughters, sisters, nieces or their cousins. Since workers also have a highly variable life span ranging from 1 to 160 days and since they do not abandon the nest or stop working after a queen replacement, intra-colony relatedness can be even more highly variable. In this long-term study, we found to our great surprise that colonies could consist of mothers, daughters, sisters, nieces, cousins, cousin's offspring, mother's cousins, mother's cousin's offspring and even mother's cousin's grand-offspring (Gadagkar *et al.*, 1993). Thus there is almost unlimited scope for strife and conflict among the members of a colony.

We therefore set out to investigate this potential intra-colony conflict. Intra-colony conflict is only possible of course if variations in intra-colony genetic relatedness can be perceived by the members of the colony. Just as the wasps can most efficiently discriminate between nestmates and non nestmates, they should be able to tell apart their sisters, from their nieces, their nieces from their cousins and so on, even though they inhabit the same nest. There was no prima facie reason to suspect that, with such well developed nestmate discrimination abilities, the wasps might be incapable of intra-colony kin recognition. And yet what we had learned about the mechanism of their nestmate recognition raised some doubts. If recognition depended on labels and templates and if naïve wasps, without appropriate labels and templates could not discriminate nestmates from non nestmates, it suggested that either labels or templates or both were acquired from the environment, in adulthood, and thus may be similar in all colony members. We coined the terms "self-produced" for labels that do not need wasps to be exposed to their nests and

11 This is the average of 100% from fathers and 50% from mothers.

12 This is the average of 100% from fathers and 50% from mothers.

"non self-produced" for labels that required such exposure. Similarly we coined the terms "self-based" for templates that do not require exposure of the wasps to their nests and "non self-based" for templates that require such exposure (Figure 8). If that was so then nestmates might be easily discriminated from non nestmates but no further discrimination of different members within a colony might be possible. In other words, all wasps could be classified as nestmates and non nestmates but once wasps mingled with each other inside a colony they would lose their identity precluding further subdivision. But of course we did not yet know for sure that both labels and templates were missing in naïve individuals.

We now set out to perform another set of experiments by recording one-way interactions between naïve and experienced wasps. For example, if experienced individuals, who should possess the appropriate labels and templates, behaved tolerantly toward naïve nestmates, then we would conclude that the naïve individuals possess the required labels and hence experience was not necessary for acquiring labels, i.e., labels are self-produced. Similarly, if naïve individuals behave tolerantly toward experienced nestmates then we would conclude that naïve individuals possessed the required templates and thus experience was not needed for acquiring templates, i.e., templates are self-based (Figure 9). In reality we found however that neither experienced individuals behaved tolerantly toward naïve nestmates nor naïve individuals behaved tolerantly toward experienced nestmates. Both interacting partners had to be experienced for tolerance to occur; even if one of them were naïve no recognition occurred. Thus we concluded that experience with one's nest in adulthood was required for the acquisition of labels as well as of templates, i.e., labels are not self-produced and templates are not self-based. This means that all individuals in a colony would have the same labels and templates and therefore no discrimination between individuals of different relatedness would be possible (Venkataraman *et al.*, 1988).

Our motivation to conduct these experiments did not originally come from an interest in studying conflict and cooperation, such an interest was aroused after we found that conflict with outside was routine and that there were mechanisms to prevent conflict with insiders (more on this below). Our motivation for these experiments came from a desire to test a rather famous theory for the evolution of social behavior and altruism. On the face of it altruism seems difficult to evolve by natural selection because rather than maximizing their chances of survival and reproduction, altruists decrease their chances of survival and reproduction in favor of others; in extreme cases they sacrifice their lives to help others to survive and reproduce. Thus altruism was deemed as an evolutionary paradox. The theory was proposed by WD Hamilton who argued that self sacrificing altruism was no paradox if we consider the point of view of genes. If the bearer of an altruistic gene kills himself and aids the survival of more similarly altruistic individuals, then the act of sacrifice will actually result in the increase in the survival of altruistic genes, not withstanding the loss of a few altruistic individuals. This has come to be known as inclusive fitness theory or kin selection theory. See Gadagkar (1995) for a review. It follows then that altruism should be observed to be preferentially directed toward close genetic relatives who have a high probability of sharing genes with the altruist. If an individual gives up reproduction, as sterile workers in insect societies do, then their sacrifice should aid in the spread of more copies of the altruistic gene than might have been the case if the altruist had herself reproduced. In diploid species (where everybody inherits two sets of genes, one from the father and another from their mother) this is a bit difficult because every individual is related to her offspring by ½ and by not more than ½ to her closest genetic relatives namely, her sisters.

	WASP A	*RESULT OF BEHAVIORAL ASSAY*	*WASP B*	*CONCLUSION*
I	Experienced (A)	Recognition → ← Recognition	Experienced (B)	Behavioral assay is working
II	Naive (A)	No recognition ⇢ ⇠ No recognition	Naive (B)	Either label not self-produced or Template non self-based or both
II	Experienced (A)	Recognition → ⇠ No recognition	Naive (B)	Label self-produced but Template non self-based
IV	Experienced (A)	No recognition ⇢ ← Recognition	Naive (B)	Label not self-produced but Template self-based
V	Experienced (A)	No recognition ⇢ ⇠ No recognition	Naive (B)	Label not self-produced and Template non self-based
VI	Naive (A)	Recognition → ← Recognition	Naive (B)	Label self-produced and Template self-based

Figure 8 An experimental approach to distinguish between the roles of labels and templates in kin recognition. A and B are two animals (say, wasps) who may or may not recognize each other as close genetic relatives depending on their rearing conditions. Based on this one can infer the ontogeny of the labels and the templates, i.e., whether labels are self-produced or not self-produced and whether templates are self-based or non self-based. See text for details. Reprinted with permission (Gadagkar, 1985).

The only way then to make more copies of their genes through altruism is to rear more siblings than offspring given up. This would require special conditions where altruists work harder or have greater success compared to selfish individuals. It is not easy to see why this should be so.

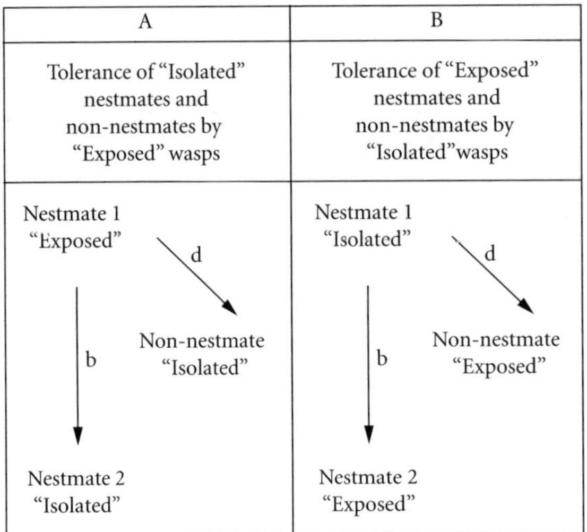

	A	B	
1	$T_b > T_d$	$T_b > T_d$	Labels self-produced and Templates self-based
2	$T_b > T_d$	$T_b = T_d$	Labels self-produced and Templates non self-based
3	$T_b = T_d$	$T_b > T_d$	Labels non self-produced and Templates self-based
4	$T_b = T_d$	$T_b = T_d$	Labels non self-produced and Templates non self-based

Figure 9 Experimental design of mixed triplets aimed at distinguishing between self-produced and non self-produced labels and self-based and non-self-based templates. In part A the ability of an "Exposed" wasp to discriminate between a nestmate and a non-nestmate both of whom are "Isolated" is assessed. Conversely, in part B the ability of an "Isolated" wasp to discriminate between a nestmate and a non-nestmate both of whom are "Exposed" is assessed. The six possible types of interactions between the three animals in these triplets are similar to those in Figure 6, but only the relevant interactions, that is, b and d, are labeled. Discrimination is again assessed by comparison of tolerance indices calculated for b and d as in the previous set of experiments. See text for details. Reprinted with permission (Venkataraman *et al.*, 1988).

In ants, bees and wasps however only the females are diploid while the males are haploid. This makes a female wasp related to her full sister by ¾ as compared to the usual relatedness of ½ with her daughter, as discussed above. Thus an altruistic sterile worker rearing sisters, needs to do less work as compared to a fertile individual who rears her own offspring, to get the same fitness. This so called haplodiploidy hypothesis can potentially explain why altruistically sterile workers are so common in ants, bees and wasps compared to other diploid organisms. However, we had shown that in *R. marginata* mating of the queen by more than one male and frequent queen replacements reduce the intra-colony relatedness so that workers were actually rearing brood related to them by less than ½. This would be a problem for the theory unless of course we could show that workers do not dispense altruism indiscriminately toward everybody in the colony but discriminate effectively between low relatedness and high relatedness values within the colony and behave in a selectively nepotistic manner. Hence we were interested in studying the possibility that intra-colony kin discrimination was possible. As it happened, we showed that intra-colony kin recognition is not possible, knocking down the haplodiploidy hypothesis. It must be emphasized that we did not knock out Hamilton's original inclusive fitness theory but only its derivative, the haplodiploidy hypothesis (Gadagkar, 2001).

Nevertheless we decided to put the possibility of intra-colony kin discrimination and the resulting potential for conflict to a further direct test. We decided to directly look for conflict inside the colony. This is not easy to do because although there is plenty of intra-colony genetic heterogeneity, it is not easy for the experimenter to discriminate individuals based on their genetic relatedness; how then can he know whether the wasps discriminate each other based on genetic relatedness? Our observation that young foreigners are accepted into alien nests gave us the possibility of creating genetically mixed nests, allowing us to mark the wasps beforehand for our easy recognition. Thus we created 12 genetically mixed colonies by introducing young, non nestmates into observation colonies in the laboratory. We took several precautions in doing this experiment. First we ensured that the observation colonies, which served as foster colonies and the donor colonies that yielded the wasps for introduction, were originally collected from well separated localities so that wasps from the two sets of colonies were unrelated to each other. Second, we made observations on the foster colonies before introducing the non nestmate wasps so that we could compare the foster colonies before and after the introduction. Third, we matched introductions of unrelated wasps carefully with eclosion of new individuals in the foster colonies so that the unrelated wasps had age-matched partners among the resident wasps, which could be treated as controls while studying the behavior of introduced wasps. In all, eighty-five relatively young wasps were introduced and all were accepted into their foster colonies, without any aggression. Thus there was no evidence of intra-colony kin discrimination. We compared the behavioral profiles of pairs of introduced and resident wasps and failed to find any significant difference; nor was there any difference in the rates of behavioral interactions between kin and non-kin wasps. Social organization, division of labor and cooperation were unaffected in the genetically mixed colonies. Even more striking was the result that some of the manually introduced wasps went on to become queens of their foster colonies. We concluded from these experiments that the genetically unrelated intruder wasps were not only accepted into their foster colonies but became well integrated and behaviorally indistinguishable from the resident wasps, eventually becoming foragers and even having a fair chance of becoming replacement queens (Arathi *et al.*, 1997b).

The results of this experiment reinforced the idea that the advantage of altruistically working for a colony for some time and eventually taking over a reproductive role to fulfill selfish interests, may well be one of the factors that drive the evolution of altruistic worker behavior in this species. Be that as it may, the striking aggression toward non nestmates as long as they belonged to foreign colonies and the striking lack of aggression when former non nestmates joined one's own colony was a remarkable and quite an unexpected result – unexpected based on all the impressions we had thus far gathered about the wasps and unexpected based on the theory of kin selection that had been proposed to explain the evolution by natural selection, of social behavior and altruism in such insect societies.

Queen succession

That the wasps were so ready to engage in war with outsiders and so unwilling to wage war with insiders was intriguing enough to motivate us to re-examine the issue in more detail. Perhaps we had been previously focusing on a situation where, despite the presence of genetic heterogeneity in the colony, there really was not that much scope for conflict. From an evolutionary point of view the main reason for conflict of course is unequal reproduction. Was it therefore possible that conflict erupts only at the time when the old queen is to be replaced by a new one? We now explored this most conflict-prone situation in natural colonies which as we saw above already have a great deal of genetic heterogeneity. To do so we designed a different kind of experiment. We studied normal colonies before any manipulation, identified the queen, experimentally removed her, studied the queen-less colonies and later studied the colonies again after returning the queen. The result of removing the queen was most spectacular. Within minutes of removing the queen the reasonably peaceful colony became a highly aggressive one. There was a several-fold increase in aggression compared to the time when the queen was present. Naturally we thought that we had solved the puzzle of intra-colony peace seen at times other than queen replacement. But the elevated aggression seen after queen removal turned out to be a red herring. All of the elevated aggression was shown only by one individual and it was all unidirectional. The aggressor was never challenged; she did not receive any aggression from any of the others. This was hardly an expression of conflict; it seemed more like take over by a pre-decided successor. That the aggressor was indeed the successor was easy to verify because if we did not return the original queen the aggressor went on to become the next queen within about a week. We therefore labeled the aggressor "potential queen" until she actually began to lay eggs when of course she would be properly entitled to the label queen. The result of returning the queen was equally interesting. The hitherto highly aggressive potential queen dropped her aggression and went back to being a normal worker as soon as we replaced the queen. The queen was unchallenged by the potential queen or by any other colony member. The potential queen, who clearly accepted the superior status of the queen was in turn unchallenged by any other wasp in the colony, irrespective of the presence of the queen (Premnath *et al.*, 1995; Premnath *et al.*, 1996; Sumana and Gadagkar, 2003). We seemed to be back to the idea of intra-colony peace and an orderly transfer of power without an overt expression of conflict.

This fascinating result deserved to be probed further. Every time we removed a queen we could identify a potential queen due to her hyper-aggressive behavior but we were unable to identify the potential queen before removing the queen. Naturally the identity of the queen's successor and the mechanism by which she is chosen is of great interest. We undertook a number of specific studies with the express goal of identifying the potential queen before removing the queen, but without success. We studied various properties of all the workers before removing the queen and attempted to identify some unique character of the individual who became the potential queen. But she was not unique in any way we could tell. The potential queen was not the largest or smallest, not the oldest or youngest, not the most aggressive or least aggressive, not even the one with the best developed ovaries. To this day we have been unable to predict the potential queen with any degree of accuracy and yet we can identify her without fail as soon we remove the queen (Deshpande *et al.*, 2006). The image of a society, which wages war with outsiders and maintains peace with insiders at all costs, was only reinforced by these new experiments. Why the potential queen who was not challenged by anyone else, was herself so aggressive to others however was a bit of a mystery. We first thought that she might beat up everybody in the colony in order to pre-empt anyone else from attempting to become a potential queen. This was probably an unlikely hypothesis because if this were so I would expect the potential queen to have received some challenge. We have now solved this mystery. We have found evidence that the potential queen needs to show all that aggression in order to facilitate the rapid development of her own ovaries so that she could start laying eggs quickly. We have shown that a potential queen without the opportunity to show aggression takes longer to develop her eggs as compared to one that has such opportunities (Lamba *et al.*, 2007). We think of this in analogy with exercise physiology and suspect that expression of aggression results in physiological changes in the aggressor so as to permit her to develop her ovaries rapidly (Gadagkar, 2009a).

I have since got over the initial frustration at not being able to predict the potential queen and now I find that inability quite charming; being intellectually defeated by the wasp is somehow pleasing. I will describe one more set of experiments that have given a new, even more charming twist to this tale. Some years ago we wondered whether the wasps knew who the successor would be even though we did not. Although we cannot predict the identity of the potential queen (hereafter, PQ), the facts that (i) only one individual steps up her aggression after queen removal, (ii) the swiftness with which she does so, and (iii) she is unanimously accepted by the rest of the workers, led us to suspect that there might be a pre-designated successor to the queen who may be "cryptic" to us in the presence of the queen but known to the wasps. At first, whether the wasps knew the identity of the successor seemed like a question that cannot be answered, for how would we know the mind of a wasp? We have now performed a most intricate experiment that does in fact permit us to answer this question. Some years ago and for a different reason, we had designed a so-called mesh experiment. Instead of removing the queen altogether we had found a way of separating some of the workers from the queen. We moved a colony into a cage, cut the nest in half and separated the two halves with a wire mesh screen and released the queen on one side – the queen-right side. We then randomly introduced half the workers on the queen-right side and the remaining half on the queen-less side. The wasps in the two sides could not interact with each other although any chemicals could flow through. We had reason to believe that the queen used a pheromone to signal her presence to the workers so that no potential queen emerged in her presence and we wanted to determine whether the queen pheromone was volatile or not. We found that the queen pheromone was non-volatile because while the queen-right side remained peaceful, the queen-less side became hyper-aggressive with a potential queen who also went on to become a queen on her side, if the wire mesh partition was not removed (Sumana *et al.*, 2008). Later we found this experimental set-up and its result very handy.

A modification of the mesh experiment permitted us to design an experiment to test the "cryptic successor, known to the wasps" hypothesis. The modification was that after a PQ became evident on the queen-less side of the wire mesh, we exchanged the PQ and the queen from side to side, leaving the workers undisturbed. The logic of this queen-PQ exchange experiment is as follows. Because the workers are randomly distributed between the 2 sides, the primary cryptic successor, if there is indeed one, has a 50% chance of being on the queen-right and 50% chance of being on the queen-less side. In those experiments where the cryptic successor happened to be in the queen-less side, she would become a PQ (we call her PQ1) and, being the true successor, she should be acceptable to the workers on both sides even when she is moved from side to side.

And in those experiments where the primary cryptic successor happened to be on the queen-right side just by chance, a different individual should become the PQ1 on the queen-less side because of the absence of the legitimate successor on her side. But this PQ1 should be unacceptable to the workers on the opposite side when she is moved to there. Instead, the real cryptic successor should now become the new PQ (we call her PQ2). Finally, the PQ2 should remain unchallenged on both sides when she is moved from side to side. Thus, the PQ1 should be acceptable to the workers on both sides in about half the experiments and the PQ2 but not the PQ1 should be acceptable to workers on both sides in the remaining half of the experiments.

Clever as it is, this is a very difficult experiment to perform. With much difficulty, we have now managed to perform it 8 times. In 3 of these experiments the first PQ was accepted on both sides and in the remaining 5 experiments, only the second PQ was accepted on both sides (Figure 10). We therefore conclude that there is indeed a designated successor to the queen. But we refer to her as a cryptic successor because we cannot identify her in the presence of the queen. An important feature of our results was that neither PQ1 nor PQ2 ever received a single act of aggression from any individual, although they themselves showed high levels of aggression. Thus, when we say PQ1 was unacceptable when we moved her to the opposite side, we simply mean that she, on her own, stopped being aggressive and went back to work although she was never challenged by any wasp, not even by the PQ2. Hence we argue that the cryptic successor is "known" to the wasps even though we cannot identify her in the presence of the original queen (Bhadra and Gadagkar, 2008). I must confess that we have no idea how the decision about who should be the successor to the queen is made. But the point I wish to emphasize here is that the decision is made before the loss of the original queen so that there is no overt conflict when the queen dies. That all the wasps know who the successor would be in the event of the queen's death, helps to ensure that power is transferred from one queen to the next without a breakout of internal conflict. Furthermore the decision appears to be made peacefully at whatever time it is made because we see no overt aggression before removing the queen. It is most remarkable that the wasps do not display intra-colony aggression and conflict, even in the context of queen succession. Indeed, we would like to create and study situations where there is conflict within a colony but have so far failed. War with outsiders and peace with insiders appears to be the invariant strategy of the wasps. I must mention that the strong nestmate discrimination abilities we have documented in *R. marginata* are not unusual; the same is true of most insect societies that have been studied (Hölldober and Wilson, 2009). But *R. marginata* is perhaps most extreme in the striking lack of intra-colony conflict and especially in the peaceful transfer of power from one queen to the next.

In retrospect, the propensity to make war with outsiders and maintain peace with insiders should not be so surprising after all. I believe that it is this dual strategy, this ability to tread a fine balance between conflict and cooperation, that accounts for the success of insect societies. Of course the war with outsiders is easier to explain than the peace

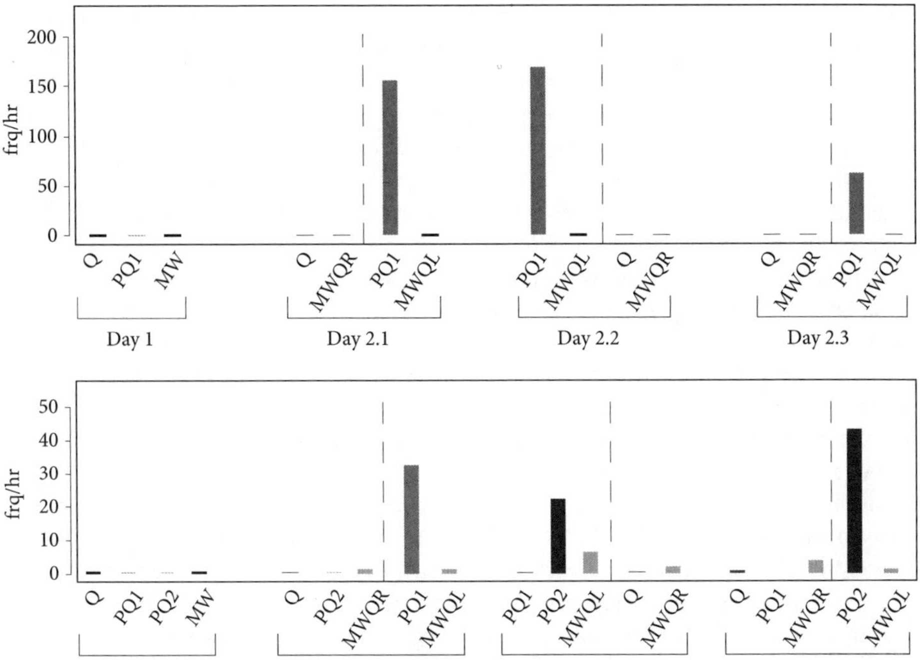

Figure 10 The Q-PQ exchange experiment designed to check if the wasps know the identity of the successor to the queen. Upper panel: A typical experiment in which the PQ1 was the cryptic successor. The frequency per hour of dominance behavior exhibited by the Queen, PQ1 and Max worker (defined as the worker showing maximum aggression) on day 1 in the normal colony, and on the queen-right and queen-less fragments in the three sessions on day 2 are shown. Lower panel: A typical experiment in which the PQ2 was the cryptic successor. The frequency per hour of dominance behavior exhibited by the Queen, PQ1, PQ2 and Max worker on day 1 in the normal colony, and on the queen-right and queen-less fragments in the three sessions on day 2 are shown. See text for details. Reprinted with permission (Bhadra and Gadagkar, 2008).

with insiders. But war with outsiders is not of much use unless one can combine it with peace with insiders. Besides, the relative lack of intra-colony conflict is consistent with the recently resurrected fashion of treating insect colonies as super-organisms. It is also not really inconsistent with old fashioned individual and kin selection either, if one remembers that Hamilton's inclusive fitness theory, more precisely referred to as Hamilton's Rule, see (Gadagkar, 1997), postulated that the evolution of altruism is modulated by a balance between cost, benefit and relatedness, rather than merely by relatedness (Gadagkar, 2001).

Concluding Remarks

I am often asked why I study insect societies. It is not difficult to justify on several levels. Social insects provide excellent model systems for understanding animal physiology, pharmacology, genetics, development and above all, evolution. Some social insects such as honeybees and their relatives are of great economic importance on account of their pollination services and are the backbone of multi-billion dollar fruit industries.

Other social insects such as fire ants and leafcutter ants are serious pests of agriculture and other human activities leading to losses of like amounts of money. In recent times the study of communication and division of labor in insect societies have found profound practical applications by providing novel algorithms useful in computer science, telecommunication and work organization in industry (Gadagkar, 2009b). And yet my own motivation for studying insect societies is rather different. My motivation is similar to that of an anthropologist. As social beings we are conscious of the enormous benefits we derive from cooperation and division of labor, but we are also justifiably obsessed with the problems that social life inevitably brings with it, the potential for selfishness and conflict. A reasonable way to reflect on these issues, and indeed to understand why we behave as we do, is to turn to other societies. Anthropologists can offer us a glimpse into the lives and mores of "primitive" and "exotic" human societies. Biologists can do much more; they can offer us insights from a whole range of animal societies with millions of years of evolutionary history. And those of us who study insect societies can hope to harness wisdom from an altogether different sub-kingdom of animal life. I certainly do not think we should imitate insect societies blindly, but I do think that they can hold a mirror to us and offer us a means to reflect on our own society and learn more about ourselves (Gadagkar, 2009a). Just as we often see the way we have arranged the furniture in our house in a new light after visiting a neighbor, or see our culture in a new perspective after visiting a foreign country, a knowledge of how other social beings conduct themselves in situations that we often find ourselves in, provides new opportunities for understanding ourselves. With the message of this chapter that the wasps readily wage war with strangers and bend over backwards to avoid conflict with those with whom they live, who can escape reflecting on how we humans manage our dual proclivities for cooperation and conflict?

As school children in India we most passionately sang a song entitled "Taraanaa-e-hindi" (song of the Indians) whose opening lines were "Saare Jahan Se Achcha" (better than all lands) and which had the thumping words "Hindi hain hum, vatan hain hindostaan hamaara" (we are Indians, India is our homeland). I knew that this song was written by Sir Muhammad Iqbal, the great Urdu poet philosopher of the sub-continent who is now officially recognized as Pakistan's national poet and who continues to be much admired in India. What I did not know until recently was that the same poet later wrote another very similar song entitled "Taraanaa-e-millee" (song of the Muslim community), now sung in Pakistan, with the corresponding thumping words "Muslim hain hum vatan hain saara jahan hamaara" (we are Muslims, the whole world is our homeland). Our reference points may change and the "insiders" and "outsiders" may change but much the same song seems to be sufficient to elicit the passion of patriotism! More reflections such as these are best left to the readers for each will have his or her own unique perspective. That is the purpose of this chapter and indeed one of the purposes of such studies. Just so that we don't get completely lost in all the fascinating details of the insect world, I have attempted to prime the reader toward such reflection with the opening quotation by the 18th century French enlightenment writer of the pen name Voltaire: "It is lamentable, that to be a good patriot one must become the enemy of the rest of mankind". I specifically chose the quotation among many possibilities because I sometimes think that it is not so much how we behave, but how we lament about how we behave, that gives a deep insight into ourselves. We do not know if the wasps lament about how they behave, but we must acknowledge that they know a thing or two about cooperation and conflict and I think it is worth our while to reflect on these issues in relation to ourselves.

Acknowledgements

The work I have described in this essay was done over a 20-year period and was only possible due to the friendship and collaboration of a large number of students and colleagues including Arathi,H.S, Anindita Bhadra, Seetha Bhagavan, Nadia Bruyndonckx, Swarnalatha Chandran, Krishnappa Chandrashekara, Sujata Deshpande, Yasmin Claire Kazi, Shakti Lamba, K.Muralidharan, Meghana Natesh, Padmini Nair, Sudha Premnath, M.S.Shaila, Mallikarjun Shakarad, Anindya Sinha, Annagiri Sumana, Martin Surbeck, Arun Venkataraman. My research has been supported by the following agencies of the Government of India namely, Department of Science and Technology, Department of Biotechnology, Ministry of Environment and Forests, Council of Scientific and Industrial Research. The Wissenschaftskolleg zu Berlin provided an intellectually stimulating atmosphere for writing this essay. Here I have drawn on some passages from my The Social Biology of *Ropalidia marginata*: Toward Understanding the Evolution of Eusociality. Cambridge, Massachusetts: Harvard University Press and my Interrogating an insect society. Proc. Natl. Acad. Sci. USA, 106, 10407-10414, with little or no modification, sometimes deliberately and I suspect, sometimes subconsciously.

References

Arathi, H. S., Shakarad, M. and Gadagkar, R. (1997a) "Factors affecting the acceptance of alien conspecifics on nests of the primitively eusocial wasp", *Ropalidia marginata* (Hymenoptera: Vespidae). *J. Insect Behav.* 10, p. 343-353.

Arathi, H. S., Shakarad, M. and Gadagkar, R. (1997b) "Social organisation in experimentally assembled colonies of *Ropalidia marginata*: comparison of introduced and natal wasps", *Insectes soc.* 44, p. 139-146.

Bhadra, A. and Gadagkar, R. (2008) "We know that the wasps 'know': cryptic successors to the queen in *Ropalidia marginata*", *Biol Lett.*, 4, p. 634-637.

Deshpande, S. A., Sumana, A., Surbeck, M. and Gadagkar, R. (2006) "Wasp who would be queen: A comparative study of two primitively eusocial species", *Curr Sci,* 91, p. 332-336.

Gadagkar, R. (1985) "Kin recognition in social insects and other animals: A review of recent findings and a consideration of their relevance for the theory of kin selection", *Proc. Indian Acad. Sci. (Anim. Sci.)*, 94, p. 587-621.

Gadagkar, R. (1997) *Survival Strategies: Cooperation and Conflict in Animal Societies*, Cambridge, Mass.; Hyderabad, India, Harvard University Press and Universities Press.

Gadagkar, R. (2001) *The Social Biology of Ropalidia marginata: Toward Understanding the Evolution of Eusociality*, Cambridge, Massachusetts, Harvard University Press.

Gadagkar, R. (2009a) "Interrogating an insect society", *Proc. Natl. Acad. Sci. USA,* 106, p. 10407-10414.

Gadagkar, R. (2009b) "What can we learn from insect societies?", in, *Nature and Culture* (Eds, Narasimha R. and Menon S.), New Delhi, CSS & PHISPC. *in press.*

Gadagkar, R., Chandrashekara, K., Chandran, S. and Bhagavan, S. (1993) "Serial polygyny in the primitively eusocial wasp *Ropalidia marginata:* implications for the evolution of sociality", in, *Queen Number and Sociality in Insects* (Ed, Keller, L.), Oxford, Oxford University Press, p. 188-214.

Hölldobler, B. and Wilson, E. O. (2009) *The Superorganism – The Beauty, Elegance, and Strangeness of Insect Societies*, New York, London, W.W. Norton & Company.

Keller, L. and Nonacs, P. (1993) "The role of queen pheromones in social insects: queen control or queen signal?", *Anim Behav,* 45, p. 787-794.

Lamba, S., Kazi, Y. C., Deshpande, S., Natesh, M., Bhadra, A. and Gadagkar, R. (2007) A possible novel function of dominance behaviour in queen-less colonies of the primitively eusocial wasp *Ropalidia marginata. Behav Processes,* 74, p. 351-356.

Muralidharan, K., Shaila, M. S. and Gadagkar, R. (1986) "Evidence for multiple mating in the primitively eusocial wasp *Ropalidia marginata* (Lep.) (Hymenoptera: Vespidae)", *Journal of Genetics,* 65, p. 153-158.

Premnath, S., Sinha, A. and Gadagkar, R. (1995) "Regulation of worker activity in a primitively eusocial wasp, *Ropalidia marginata*", *Behav Ecol,* 6, p. 117-123.

Premnath, S., Sinha, A. and Gadagkar, R. (1996) "Dominance relationships in the establishment of reproductive division of labour in a primitively eusocial wasp (*Ropalidia marginata*)", *Behav Ecol Sociobiol,* 39, p. 125-132.

Sumana, A. and Gadagkar, R. (2003) "*Ropalidia marginata* – a primitively eusocial wasp society headed by behaviourally non-dominant queens", *Curr Sci,* 84, p. 1464-1468.

Sumana, A., Deshpande, S. A., Bhadra, A. and Gadagkar, R. (2008) "Workers of the primitively eusocial wasp *Ropalidia marginata* do not perceive their queen across a wire mesh partition", *J Ethol,* 26, p. 207-212.

Venkataraman, A. B. and Gadagkar, R. (1993) "Differential aggression towards alien conspecifics in a primitively eusocial wasp", *Curr Sci,* 64, p. 601-603.

Venkataraman, A. B. and Gadagkar, R. (1995) "Age-specific acceptance of unrelated conspecifics on nests of the primitively eusocial wasp, *Ropalidia marginata*", *Proc Indian natl Sci Acad,* B61, p. 299-314.

Venkataraman, A. B., Swarnalatha, V. B., Nair, P. and Gadagkar, R. (1988) The mechanism of nestmate discrimination in the tropical social wasp *Ropalidia marginata* and its implications for the evolution of sociality, *Behav Ecol Sociobiol,* 23, p. 271-279.

Wilson, E. O. (1971) *The Insect Societies.* Cambridge, Massachusetts, USA, The Belknap Press of Harvard University Press.

Wilson, E. O. (1990) *Success and Dominance in Ecosystems: The Case of the Social Insects.* Germany, Ecology Institute, Nordbünte.

2.2 Towards Understanding Conflict Elicitation and Conflict Resolution: Challenges for Brain, Behavioral and Social Sciences

by Urs Luterbacher, Sebastian Di Iorio and Beth Krasna[1]

The Geneva Lakeshore region offers a unique constellation of academic institutions that have a strong tradition in understanding human behavior and international organizations that have a major role in regulating human societies. This close regional proximity and the growing cross-fertilization between these different fields of expertise make it particularly pertinent and timely to envision novel avenues of collaboration and research on global issues to benefit mankind and society. The consortium for the Interdisciplinary Study of Conflict And Negotiation (CISCAN)[2] brings together actors from different spheres of knowledge to initiate innovative research strategies for interdisciplinary approaches to human conflict and conflict resolution, through the analysis of individual and group behavior from a multiple brain, psychological, social and political perspective, incorporating neuroscience, computational and engineering approaches. We believe that conflict elicitation and conflict resolution play a major role in the regulation of interpersonal and

1 Urs Leuterbacher is a Professor of Political Science in the Department of Political science and Chairman of the Environmental Studies Unit, The Graduate Institute of International Studies (IHEID), Geneva, Sebastien Di Lorio obtained a Masters in Political Science from IHEID in 2006. Beth Krasna holds an MS in Chemical Engineering from the ETHZ, and an MBA from the Sloan School at MIT. She is currently a member of the Conseil des Ecoles Polytechniques Fédérales.

2 This informal consortium was created to apply the interdisciplinary research done in the Geneva Lakeshore Region to conflict elicitation and conflict resolution. Academic institutions currently participating are EPFL (Ecole Polytechnique Fédérale de Lausanne; IHEID (Graduate Institute of International Studies); UNIGE (University of Geneva); and UNIL (University of Lausanne). The following people have been involved in the elaboration of the CISCAN concept: Klaus Scherer, David Sander, Benoit Bediou, Swiss Center for Affective Sciences & Psychology Dept. UNIGE; Patrick Vuilleumier, Center for Neuroscience & Neurology Dept. UNIGE; Urs Luterbacher, Political Science Unit, IHEID; Pierre Magistretti, Brain Mind Institute, EPFL, UNIL; Olaf Blanke, Brain Mind Institute, EPFL, University Hospital of Geneva, UNIGE; Philippe Moreillon, Fundamental Microbiology Department, UNIL; Beth Krasna, independent facilitator.

international relations, which can be tractable and experimentally modeled by such interdisciplinary approaches.

One of the main challenges for interdisciplinary research endeavors under an "umbrella" theme is getting the different projects to coordinate and share their research results. Some interdisciplinary projects are trying to do this by exploring common ground. Industry has solved this problem by defining an overriding goal such as "put a man on the moon", where all results eventually converge in order to achieve the main objective.

CISCAN aims to achieve convergence of diverse research outputs using dynamic modeling to incorporate them into one large interactive simulation, which will allow for increased complexity to be explored as more fields and research results get added. This approach should eventually allow research projects to add new variables and go through further iterations, thus improving the model's approximation to reality over time. These types of models are already showing themselves to be powerful aids in understanding complex issues such as climate change. The CISCAN model will be built along the lines of a "SIM City[3]" type of game, developed by physicists, mathematicians and game developers to simulate the propagation and stop/start phenomena of conflict. The notion of game is interpreted here as a stylized model of reality (such as in game theory) that allows for the simulation and thus "playing out" of different present, past, and future alternatives and not as a form of entertainment. It is hoped that a formulation of this type might actually have predictive value and might eventually contribute to the prevention or rapid resolution of conflict situations.

Conflict over scarce resources (territory, mates, food) between members of the same species is a universal feature of evolution. Force is not the only decisive factor; perceived entitlement and powerful feelings of injustice thereby generated in the case of challenge, extended to group identity are at the basis of conflict and aggression in humans. The relationship between environment and conflict, the role of emotions such as fear, and the absence or weakness of clear definition and enforcement of property rights within societies are also factors in the development of conflict and sometimes genocide. The development of weapons and complex forms of social organization allowing group cooperation in aggression have turned simple dominance contests into a spiral of ever more complex conflict mechanisms that are exceedingly difficult to control.

CISCAN's aim is to create a platform to evaluate, conduct and integrate research programs at different levels of analysis, including brain, physiological, cognitive-affective and computational mechanisms, engineering based approaches and strategic analysis to;

1. better understand the determinants and dynamics of conflict among individuals, groups, corporations and nations
2. explore the psychological, strategic and social factors involved in the process of negotiation and mediation
3. model and develop engineering-based strategies for the study of interference with social interactions
4. elaborate in close partnership with expert practitioners in the area of conflict regulation and negotiation (politicians, lawyers, diplomats, mediators, educators,

3 A city-building simulation game that spawned a series of other computer games based on principles of emergent phenomena. E.g see http://en.wikipedia.org/wiki/SimCity for more details.

managers, union representatives, etc...) recommendations for practical action and procedures based on established research findings

The academic institutions around the Geneva Lakeshore region are particularly well suited to this enterprise given;

1. the many centers of excellence in research in the institutions
2. the leading edge research capacity in the neurosciences ranging from cellular to systems and cognitive neuroscience and neuro-imaging
3. the special competence and critical mass in affective sciences
4. cutting edge expertise in virtual reality technology and animation of complex social situations and technologies for the determination of the physiological responses associated with emotional processing
5. the presence of leading international organizations in the Geneva area and the frequency of conflict resolution and negotiation efforts conducted in this context
6. existing links between academic institutions on one hand and international organizations, political institutions on a local and national level on the other
7. past record and current efforts toward interdisciplinary work

The first project to be investigated in terms of the framework we have defined is entitled:

Interpersonal and international conflict:
challenges for brain, behavioral and social sciences
towards understanding conflict elicitation and conflict resolution

conducted by Urs Luterbacher with the collaboration of Sebastian Di Iorio (IHEID).

In the following pages, the vexing question of the relationship between environment and conflict and the role certain emotions like fear play in it is explored. In the social and especially political science literature on conflict, the integration between emotional and rational choice explanations of conflict is poorly realized. Superficially, these two factors, emotions and reason seem to be at opposite ends of an explanatory spectrum. However, a common ground can be found if one follows the intuition that emotions are associated with rare and unusual events such as serious threats to existence or the prospect of severe losses of wealth or physical, property, territory or physical integrity. Such considerations are important if one wants to model decision-making behavior accurately in a simulation model. The model is described in mathematical terms in the following pages. However, it is possible for the reader to follow the overall line of the arguments without recourse to the equations.

Rational choice, particularly expected utility approaches have been poor at dealing with such situations. Early axiomatizations of expected utility theory such as those put forward by von Neumann and Morgenstern (1947) or Friedman and Savage (1948), essentially ignored the issue. It was the merit of the French economist Maurice Allais (1953) to have raised the problem of taking into account rare or very constraining events. He illustrated his criticism of the traditional approaches with the following anecdote: If a traveler is caught in Marseilles with a minimal amount of money not sufficient to buy a ticket and has to reach Paris imperatively, any gamble that will give him the greatest probability of winning more than the price of his ticket will be worth entering into regardless of absolute utility maximization procedures. Allais' considerations led to the formulation of the well-known Allais paradox in the literature. To take these issues into account Chichilnisky (2006)[4] has developed a new axiomatic system. The psychologists Kahneman and Tversky (1979) through their development of "prospect theory" have shown that Allais' considerations were backed by empirical observations and that decisions in uncertain situations are taken under two general perspectives: 1. Ordinary individuals tend to be risk averse under the prospect of gains and risk preferring under the prospect of losses. 2. They tend to overestimate the likelihood of low probability events and underestimate the likelihood of high probability ones. Fearful reactions are often motivated by the prospect of losses associated with some specific event either natural or social: A natural catastrophe or a massive social upheaval will lead to risky gambling and make conflict outcomes more likely despite their own prospective high costs.

Environmental change is often caused by a combination of natural and social forces. Human activities can lead to overuse of natural resources which then result in scarcities and these scarcities can then lead to emotional reactions and thus to taking risky gambles and to conflict. Given the fact that the empirical evidence about the relations between environment and conflict is ambiguous, it suggests that the link between the two issues only makes sense and works whenever institutional factors such as the clear definition and enforcement of property rights are absent or weak within or across societies and thus when fear about property losses motivates individuals to act aggressively. Below, the empirical case of Rwanda is used to illustrate this relationship. After a discussion of the data problems that the case raises, a simulation of the conflict and the genocide of 1994 is proposed. Though simplifying and stylizing the intricate social and political behavior to a considerable degree, the simulation model accounts quite well for the observed conflict and genocide evolution.

The Rwandan Genocide

In the Rwandan genocide of 1994, between 500,000 and 800,000 people, Tutsis and moderate Hutu members of the two ethnic groups in the country, were massacred in the space of four months. It stands out as one of the most tragic events in the history of the late 20th century. Like for other genocides such as the Shoah, this tragic event has led to a considerable amount of research as well as to various attempts at analyzing why and how it happened. In addition to the historical style representations a few formal approaches have been presented. Among those, the contribution by Bhavnani and Backer (1999) appears to be particularly elaborate because they are trying to build on previous formulations developed by political scientists on ethnic conflict. Earlier work such as that by by Fearon and Laitin (1996) and Lohman (1993) is centered on information questions. Conflict will be initiated or amplified by information circulating mostly within one group and directed against the other. Misperceptions about the other group can lead to armed

---

4 She suggests an alternative axiomatization of utility theory in order to account for attitudes involving fear of catastrophes.

violence. Without denying the importance of information questions in the generation of armed conflict, one can nevertheless legitimately ask about the importance of other factors, especially in the context of Rwanda and of Africa in general. Various authors (Andre and Plateau, 1998; Homer-Dixon, 1994; Collier and Hoeffler, 2000) have emphasized the importance of resource issues, at least as a long-term factor in explaining conflict. The information issues raised by Bhavnani and Backer (1999) are raised without reference to a specific context. However, context matters and in the case of the Rwandan genocide as in others, the emotion of fear appears to be a powerful motivator for extreme behavioral responses in general and for conflict and violence in particular. Quite clearly, fear is also closely linked to rumor and thus often to false information transmissions. Examples of false rumors generating fear among populations are numerous. A telling example of such rumor spreading is the episode known historically as the Great Fear (*La grande peur*), which overtook much of the French country side in the late summer and early fall of 1789 and during which time peasants feared armed interventions by angry aristocrats and their partisans (Lefebvre, 1973). These fears turned out to be completely groundless but they could have triggered violent reactions. The Great Fear incident illustrates how peasants apprehended the possible reactions of the aristocratic minority which was still associated with military prowess and thus perceived to harbor a desire for revenge at being stripped of all privileges by the ongoing revolution.

We will claim that similar emotions were at work in Rwanda. Rwanda has had a history of conflictual relations between the alleged Tutsi and Hutu ethnic groups (even though the distinction and boundary between them is incredibly fuzzy) even before independence in 1962. The Tutsi king and ruling class were driven from power in 1959 by Hutu forces which then managed to expel some Tutsis and to put in place and maintain a Hutu dominated government which lasted until 1994. The Hutu government was able initially to expand the land base and thus to raise agricultural production in part because the expulsion of some Tutsis made more land available for cultivation. But this action had another more detrimental effect as expelled Tutsis formed a guerrilla movement outside of Rwanda which then intervened more and more frequently within Rwanda itself, harassing the Rwandan governmental army. Eventually, the Rwandan government was unable to keep agricultural production growing while the Rwandan population kept expanding and shortages started to appear. The resulting discontent gave rise to further attempts at encroachments on Tutsi land and thus indirectly strengthened the guerrilla movement. As the guerrilla movement grew in force threatening to overpower the Rwandan army, fear grew not only that the Tutsis would take over as they had in the past, but also that they would seek revenge for the years of discrimination and expulsions. We suggest that it was essentially this fear which triggered the bloody events of 1994, a desperate attempt by the prevailing Hutu powerbase to solve the Tutsi question once and for all. Despite its horror, however, the massacre of hundred of thousands of Tutsis and moderate Hutus failed in its goal: The Tutsi-led guerrilla prevailed and took over, but only after one of the most horrific massacres of the 20th century had taken place. As this narrative of the historical events emphasizes, the outbreak of violence triggered by fear took place amidst growing scarcities of land and resources. This particular factor will be examined first.

Environment Based Conflict

Are environmental scarcities responsible for conflict? This vexing question has haunted the literature since Thomas Homer-Dixon rekindled it in 1994. The answers however have been far from unambiguous because various theories and even empirical analyses invoke several, often opposing relations: people fight because too many are competing for too few or degrading resources (Homer-Dixon, 1994; Diamond, 1997, 2004) or, conversely, because some groups have unfettered access to natural resources, which allows them to finance rebellions (Collier and Hoeffler, 2000; Journal of Conflict Resolution, 2005). Such opposing views suggest that if a direct causal relationship between environment and conflict exists at all, it is complex and involves many factors (Gleditsch and Urdal, 2002). In this paper, we will try to examine such complexities and see if the two perspectives can be put under a common framework.

This common framework can be found in institutional structures. Institutional factors leading to resource scarcities gained prominent attention through the "tragedy of the commons" argument put forward by Garrett Hardin (Hardin, 1968). He suggested that the nature of property regimes directly influenced resource use and hypothesized that some forms of collective management created incentives to dissipate the resource. He did not directly address the problem of conflict but showed that the positive feedbacks engendered by the absence of well-defined property rules should lead to scarcities. These could trigger desires to appropriate or over-use resources at the expense of other groups or individuals. In other words, environmental scarcities need not necessarily lead to conflict, but the absence of an institutional structure that regulates these does. These issues can be analyzed first from a theoretical point of view.

Theoretical Considerations about Resources

The dynamics of open access or unregulated social systems as imagined by Harding can be conceived more precisely as the interaction between a resource stock and a population that uses it. If the resource stock is finite it will eventually be depleted. It is however possible to deplete it at an optimal rate which should allow timely switches to the use of other resources. If the resource is renewable, the particular dynamics of its evolution will have to be taken into account especially given the fact that the population will use it at a rate that will be compatible or not with its sustainability. Let us imagine that in a simple production system that relies on a renewable resource (such as fishing), the cumulative production can be represented by an S shaped curve while cumulative production costs can be expressed as a straight line if constant unit costs per individual producer are assumed. The production issue in an open access system can be illustrated by the following graph:

Figure 1 shows that up to A, marginal productivity is higher than average productivity but that the output curve is lower than the cost curve. In other words there is an initial investment to be made in terms of sunk costs in order to reap the benefits of greater outputs, which are achieved after point A. The problem, in terms of a collective good creation can be defined as how to convince (or to force) enough producers to participate especially in view of the fact that average product (which motivates producers always in a collective good situation) is lower than marginal product.

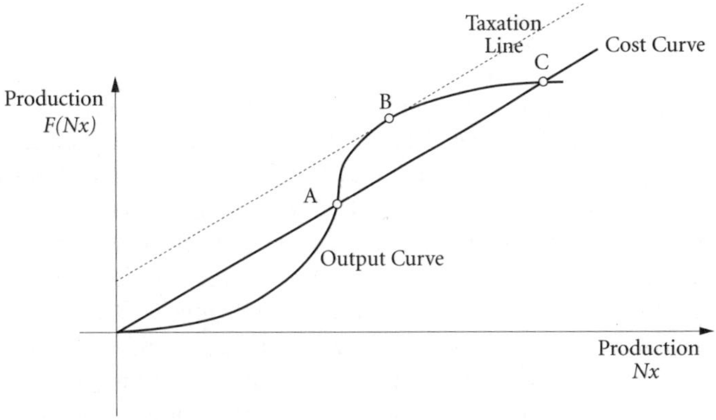

Figure 1 Production from a renewable resource in an open access system.

After point A, the problem becomes opposite in the case of Figure 1[5], Output outstrips costs and surpluses are generated which reach their maximum at point B where the slope of the output curve is equal to the marginal cost. If now more and more producers with their inputs are allowed into the process, the maximum at point B is passed, the whole surplus is dissipated at point C, and one is faced with the familiar tragedy of the commons. It can be noticed that if a tax (a Pigouvian tax)is imposed on production proportional to costs, then maximum profit can be reached at point B, which can then be interpreted as a tax equilibrium[6].

In the absence of taxation and regulation however, we can formulate the above considerations in terms of two dynamic differential equations, which describe:

1. The evolution of the resource stock with the basic assumption that it grows subject to its own natural dynamics minus what is being produced (i.e. extracted from it):

$$\frac{dz}{dt} = H(z) - F(z, Nx) \tag{3.1}$$

where z is a variable that represents the resource stock and *F(z,Nx)* a cumulative production function similar to the one in Figure 2 which takes into account the amount of productive input *x* provided by *N* users so that one has *Nx.*

2. The evolution of the number of inputs x, which is proportional to profit, defined as revenue minus cost, a dynamic that reflects exactly the process leading to surplus dissipation in Figure 2:

5 This is due to the particular shape of the output curve and the slope of the cost curve. A continuation of increasing returns after A is perfectly conceivable for a while even if the assumption of the S shaped output curve is maintained.

6 The Pigouvian tax is named after the British economist Alfred Pigou's work on taxation. The term describes a tax intended to correct for the impact of a given market transaction on parties not directly involved in the transaction itself. On the notion of a tax equilibrium refer to Dasgupta and Heal, 1979, p. 52-54.

$$\frac{dNx}{dt} = \mu(qF(z,Nx) - pNx) \qquad (3.2)^{7}$$

Where x again represents input per producer, $F(z,Nx)$, a production function, q the price of the product and p the cost of a unit of input and where : is an adjustment constant between revenue and cost in terms of additional inputs x. In other words, equation (3.2) tells us that new entrants (represented here by more inputs) will move into this productive activity as long as profits can be made. This occurs of course because no limits are placed on engaging in that activity exactly as assumed also in Figure 1. This general formulation is due to Dasgupta and Heal (1978, p. 122 and 134). Obviously if taxes corresponding to the scarcity rent of the resource and an "entry" fee to start using it are charged then the problem of overuse disappears. However, this solution requires cooperation in order to establish a resource-preserving regulation. In the absence of a regulatory framework, a Pareto suboptimal situation[8] will prevail since total profits will have been driven to zero and the scarcity rent associated with the resource, dissipated. This particular kind of Pareto suboptimal situation is equivalent to overuse. Equations (3.1) and (3.2) are in fact general forms of the Lotka-Volterra equations which describe in mathematical terms evolutions of prey and predator populations. In general if one deals with a subsistence type economy, we can consider a relatively fixed input so that we can set $x = 1$ such that only the dynamics of N, the population matter. Clearly, left to themselves these dynamics will usually lead to resource exhaustion and hence population collapse. Such population collapses are also often preceded by conflicts as for instance in the case of Easter Island[9]. We can assume that such conflicts erupt when individuals find it more attractive to appropriate by force rather than to produce. We can readily see how a combination of the dynamic formulations suggested above can account for the empirical evolutions of collapsing or severely conflict ridden societies (such as for instance Rwanda). An illustration of a conflict situation can easily be derived from the above relations. This derivation constitutes our extension of the basic Dasgupta and Heal model which we introduce in the following way: Assume that a resource stock is the object of a competition between two groups which we can designate as populations N and M. Their respective production functions based on the resource z can now be designated as $F(z, N,M)$ for population N and $G(z, M,N)$ for population M. We can assume that the productive activities of either N or M might interfere with each other (usually negatively) and thus the production functions should include the size of the other population as an input. We have then if we keep our previous assumptions:

$$\frac{dz}{dt} = H(z) - F(z,N,M) - G(z,M,N) \qquad (3.3)$$

7 The dynamics of equations (3.1) and (3.2) are represented more explicitly in an article by Brander and Taylor (1998) describing sustainability problems on Easter Island over time. They exhibit a long-term (low frequency) population resource cycle analogous to those suggested by Volterra (1931) Lotka or Kostizin (1937) for animal populations.

8 The concept of Pareto efficiency (named after Italian economist Vilfredo Pareto) has broad applications in economics, game theory, engineering and the social sciences. In Pareto suboptimal situations, it remains theoretically possible to change the given parameters that describe its dynamics, in such a way as to make at least one individual better off without making any other individual worse off.

9 We refer again to Brander and Taylor (1998).

And respectively,

$$\frac{dN}{dt} = \mu\left(qF(z,N,M) - pN\right) \tag{3.4}$$

And

$$\frac{dM}{dt} = \nu\left(kG(z,M,N) - sM\right) \tag{3.5}$$

We assume that v and k stand for the second population M for their speed of adjustment and their price respectively. From equation (3.3), we can now replace $G(z, M)$, respectively $F(z, N)$ by their values and introduce these into equations (3.4) and (3.5) which gives:

$$\frac{dN}{dt} = \mu\left(q\left(H(z) - \frac{dz}{dt} - G(z,M,N)\right) - pN\right) \tag{3.6}$$

And

$$\frac{dM}{dt} = \nu k(H(z) - \frac{dz}{dt} - F(z,N,M) - sM) \tag{3.7}$$

We can now make the following assumptions connected to conflict:

Let us assume that the resource stock z is changed from a variable increasing (or decreasing) quantity to a fixed amount z^*. As a result we can reinterpret $H(z^*)$ as a fixed amount of z^* available for use. Since z^* is fixed, it makes sense to posit that

$$\frac{dz}{dt} = 0$$

We are then left with the following differential equations:

$$\frac{dN}{dt} = -\mu q G\left(z^*, M, N\right) + \mu\left[qH(z^*) - pN\right] \tag{3.8}$$

And

$$\frac{dM}{dt} = -\nu k F(z^*, N, M) + \nu\left[kH(z^*) - sM\right] \tag{3.9}$$

If now $G(z^*, M,N)$ and $F(z^*,N,M)$ are Cobb Douglas type production functions: $M^\alpha N^\beta$, then (3.8) and (3.9) become:

$$\frac{dN}{dt} = -\mu q M^\alpha N^\beta + \mu\left[qH(z^*) - pN\right] \tag{3.10}$$

And

$$\frac{dM}{dt} = -\nu k N^{\delta} M^{\gamma} + \nu\left[kH(z^*) - sM\right] \tag{3.11}$$

These are generalized forms of the Lanchester (1916) equations, concentration or dispersion combat equations which describe the evolution of two population groups (or armed forces) opposed to each other in a violent confrontation. Our representation shows therefore how resource competition can degenerate into open conflict. Moreover, it also expresses the notion that combat participation might be enhanced by access to natural resources. Quite clearly, such an outcome is not inevitable. It can only happen if two conditions are met. First, if long-term considerations about the availability of resources are not internalized by agents and, second, if their entry into the competitive production process of resources is not limited. The absence or poor enforcement of regulations and well established property rights could then lead to combat and fighting. Obviously, once fighting starts, regulations and property rights systems tend to break down further and combat prevails even more. Lanchester relations are then appropriate to describe its evolution.

In particular, when $\beta = \gamma = 0, \alpha = \beta = 1$, in (3.10) and (3.11) we get a form of Lanchester's square law, where troop concentration leads to more than proportional casualties on enemy forces and when $\beta = \gamma = 1, \alpha = \beta = 1$, we get a form of Lanchester's linear law where dispersed forms of combat lead to proportional casualties on the other side. We can therefore establish how competition for a resource can lead directly to an armed conflict when the resource is finite. We should thus be able to apply some forms of the Lanchester combat equations to conflicts connected to resource scarcities. We have to remember however that these scarcities are not necessarily resulting from some natural phenomenon but more often are the result of man made overuse. We will now elaborate on the Lanchester formulations.

Theoretical Issues in Conflict Analysis: Combat Models

Any evolving violent conflict interaction can be characterized as a succession of combat situations. The Lanchester combat equations represent a convenient instrument to describe such an evolution. Lanchester (1916) tried to describe combat with the help of two central concepts, concentration and dispersion in a struggle between two opponents. In a frontal clash of two armed formations, the one that reaches the highest degree of concentration with the biggest army will win. Conversely, when combat becomes dispersed as in artillery fights, air bombings or naval battles, the force with the highest capacity to blanket an entire area with its blows has the upper hand. Lanchester buttressed his arguments through the use of mathematical descriptions of the two types of combat in the form of differential equations. So concentrated combat between two forces is described by the following equations:

$$\frac{dx}{dt} = -k_1 y \quad \frac{dy}{dt} = -k_2 x$$

Where x is the size of the armed forces of one side, y of the other side and k_1 and k_2 are kill probabilities of respectively one unit of y hitting one unit of x and vice-versa.

Dispersed combat between two armed groups is represented by the following:

$$\frac{dx}{dt} = -r_1 xy \quad \frac{dy}{dt} = -r_2 yx$$

Now r_1 and r_2 stand for the respective kill probabilities with a different meaning to be explicated below.

This mathematical apparatus is mainly descriptive in the sense that it does not include any modeling of the decision making process that involves the commanders of the armed forces under consideration. Lanchester's results are thus deterministic and entirely dependent on some parameter values such as the initial values (sizes) of each of the armed forces facing each other, the kill probabilities of the fighting units and, in the case of dispersed combat, areas targeted for hits. Nevertheless, we have tried to show above that these relations can also be deduced from increased scarcity conditions resulting from a myopic overuse of natural resources. Within the Lanchester perspective, the armed force that declines the most rapidly in case of equality of sizes will eventually be the losing one. Even though the limitations of the Lanchester approach are readily apparent, it nevertheless fulfils the important role of giving potential indications about the costs associated with warfare since it gives a rough idea about the initial size and kill probabilities needed to overcome an enemy. It can also give an indication of the dynamics of conflict and their non-linearities: Losses will be extremely numerous initially and only diminish progressively as a faction is going through warfare.

Sometimes losses will accumulate rapidly beyond replacement possibilities in ways that are difficult to anticipate. Most civil wars and especially low intensity combat or guerrilla warfare type dynamics are difficult to model using pure Lanchester concentration or dispersion relations because they usually involve mixed conditions of the two types of armed confrontations. For these reasons, Deitchman (1962) introduced a mixed concentration dispersion Lanchester model to account for guerrilla and low intensity conflicts. Deitchman presents the strategic situation of guerrilla fighters in the following way: The guerrillas are usually dispersed over a territory which forces government or occupying forces to attack them in a dispersed way for instance by blanketing a whole region with search and destroy missions, artillery fire or even massive bombings. These were tactics used by American forces in Vietnam and also to some extent now in Iraq. Guerrilla forces on the other hand, can attack targeted governmental or occupying forces in a concentrated way, which they do mostly by using ambushes. In addition, guerrilla fighters depend largely for the survival of their efforts, on the existence of a part of a population that supports them and provides with a base for recruitment purposes. There is thus a fundamental asymmetry between the guerrillas who fight in a concentrated way and the government or occupying troops that have to undertake dispersed combat operations. This situation has two important consequences. On the one hand, being forced to fight in a dispersed manner, government or occupying forces will inevitably hit civilians who have nothing to do with the guerillas and exert some form of "collective punishments". This will often turn the population that the guerrillas claim to represent even more against the government or the occupier[10]. Another way to weaken guerrilla forces is to shrink the fraction of the population that supports them through violent action up to and including genocide. We will try to stress the practical meaning of these concepts in the concrete case of Rwanda.

Empirical Analysis

These latter considerations lead to the question of the empirical evidence behind our formulations. A dynamic simulation of a resource based conflict will then be carried out. For this we chose the dramatic case of Rwanda where between 500,000 and 800,000 people were killed in 1994.

We can hypothesize that there may be a temporal sequence implied by our theoretical considerations above. In a first stage, strong demographic expansion would lead to political difficulties that in turn lead to autocratic regimes. In a second, these autocratic regimes would eventually collapse as the children produced by the demographic expansion reach adulthood and contribute to an excessive population density. The scarcities resulting from population pressure on resources could lead to civil strife, ultimately overturning the regime. Rwanda represents a case where such a sequence might have been at work. We will now analyze it.

Rwanda had a very difficult history of social and economic relations even before gaining its independence in 1962. In 1959 the Tutsi king of the country was overthrown by the majority Hutu group. Some of the Tutsi minority were then killed or forced into exile, mostly into Uganda where they formed The Rwandan Patriotic Front, a revolutionary group bent on changing the existing order in Rwanda. At first they were not successful since the government of President Juvenal Habyarimana, was able to promote agriculture, the main economic activity of the country through substantive extensions of the areas under cultivation. This extension occurred at the expense of marshes and forests, but also through the reoccupation of plots abandoned by segments of the fleeing Tutsi population. Eventually this policy reached its limits and was especially unsuccessful at checking population growth. Rwandan agriculture has always been prosperous thanks to favorable climatic and ecological conditions. Caloric production per capita increased by 22% between 1965 and 1982, only to fall back to its 1960s level in the last decade of the century (André and Platteau, 1998). To the extent that the per capita production of food crops followed the same pattern (*ibid.*), one may, with hindsight, really question the strategy set up by the Rwandese authorities. In particular, the relation linking the abundance of natural resources and the form of social and political controls it implied seems critical to understand the historical trajectory of the dramatic events that took place in 1994.

Two aspects of the policies that were put into place stand out. First, the government's strategy mainly promoted developing new land and decreasing fallow, resulting in increasing returns. The limits to such a strategy were reached as population densities eventually converged across the country, as compared to the wide disparities that prevailed until recent times (André and Platteau, 1998). Moreover, the production technology remained

[10] On the other hand, if the population attributes the blame to the guerrillas, the government's popularity could then increase.

highly traditional and faced severe problems of erosion and soil mining (due to the utilization of forested and pasture land for cultivation). The second aspect is the emphasis put on food self-sufficiency, illustrated by the fact that the country's per capita exports are among the lowest in the world (André and Platteau, 1998), proscribing the abandon of low yielding, traditional crops. Thus, in the face of a sustained population growth of well over 3% per year, it is not so surprising that famines made reappearance by the late 1980s in several areas (André and Platteau, 1998).

The indications one can find in the literature about land tenure are difficult to interpret. It seems that the main mode of land acquisition has constantly worked through inheritance (as expected in Sub-Saharan countries). The role of individual ownership and transmission from ascendant to descendant is not clear however. It seems that land was mostly, in accordance with the custom of the region, communally owned (Economic commission for Africa: 2004). Colonization appears to have attempted to introduce landed property rights similar to the Western concept but in practice the attempt was limited here, as in other sub-Saharan African countries, to European land-aquirers and big firms. As a reaction many post-colonial governments either abolished western style property rights altogether or exerted strict control over land tenure, which was the case in Rwanda. Thus, well-defined property rights were never established and the population was led to perceive the government and not individuals, as the provider of land. In fact, in Rwanda, given official policies, the government was probably seen as the provider of land of last resort, especially if more could be appropriated from weaker minority groups. Given such expectations, demographic incentives worked in the wrong direction: the population was led to expect that the possibilities to cultivate land were limitless and thus more children were produced. In accordance with Demsetz's ideas, a land market eventually developed when population growth and density led to land scarcities. This implied a wide set of consequences similar to what one would find in a black market: inequalities in access to land rose, conflicts among family members over inheritance increased dramatically, along with disputes over land. Worth noting is the fact that "many land parcels were sold under distress conditions and purchased by people with regular non-agricultural income"(André and Platteau, 1998, p. 28), which shows that those who did not have the possibility to earn additional sources of income fell into a sort of poverty trap: by selling their land they lost the ability to get out of poverty. In other words, this type of market was not really legal and was initiated much too late to check, at least in the short run, demographic expansion. In addition this black or grey form of buying and selling land implied the erosion of traditions and customary rules, because, as a commodity, it became independent of such notions. Thus, one can see that scarcities in resources have tended to magnify inequalities through (illegal) market operations.

After the assassination of president Habyarimana under murky circumstances a *coup d'Etat* was initiated to prevent further expansion of the Tutsi guerrilla army. The organizers of the *coup d'Etat* formed a small group belonging to the regime's political, military and economic elite, who at a time had once been close to the president, and whose goal was to stop any form of democratization (Prunier, 1995). While they profited from the involvement of the Presidential Guards – to the extent that it provided a highly organized group capable of targeting selected individuals and groupings – it is clear that the main agents of the genocide were the peasants themselves. As Prunier puts it, "their [the organizers] efficiency in carrying out the killings proves that these had been planned well in advance… but it would not have been enough had it not been for two other factors: the capacity to recruit fairly large numbers of people as actual killers and the moral support and approbation of a large segment – possibly a majority of the population". Thus, the costs of organizing and sustaining an uprising had been considerably lowered by (1) the scarcities in land and opportunities of off-farm income and (2) discursive strategies that served to mobilize high numbers of poor, unemployed and uneducated young men without any perspective to inherit land. It should be noted that the prize coveted by the plotters was political power, whereas peasants acted out of strong personal grievances: "all these people who were about to be killed had land and at times, cows. And somebody had to get these lands and these cows after the owners' death" (Prunier, 1995). Hence, the issue of ethnicity may be validly considered more as an instrument in the hands of decision-makers than a cause of the conflict. The underlying and ultimate reason is more likely to be found in the combination of resource scarcities and declining state power. Indeed, one should note that Hutu and Tutsi are not tribes but social groups inside the same culture (Prunier, 1995). This had allowed mixed marriages and prevented a systematic separation of dwellings in mixed population zones. Thus, people had lived together and side-by-side all the time. But André and Platteau reveal that even in the areas where you find only Hutus, there were lots of victims of the genocide: All the victims in their sample area were Hutus (since there were no Tutsis in the villages studied). This sustains the theory of resource scarcities.). The fact that "intra-ethnic" killings nevertheless took place is an indicator of the political (as opposed to ethnic) feature of the crisis.

Simulating the Conflict and Genocide

To summarize the scenario suggested by this historical narrative, we can say that the conditions setup at independence led to expectations of increased land availability either through appropriations from minority groups or through gain from marsh draining and deforestation. As a result, birth rates exploded and a demographic expansion took place. These trends are illustrated in the following graphs. Figure 2 shows the increase in available arable land as the Rwandan government cleared marshes and forests to expand the total area. However this expansion came to an end in the late eighties and even began to reverse in the early nineties.

This was paralleled by a demographic expansion, which led to a population increase as well as the persistence of a high population growth rate until 1994. The population density of rural areas also increased considerably from 1970 on.

Given these trends and the kinds of incentives that prevailed, land resources were eventually all used up and a violent confrontation between two competing groups, which can be described in terms of Lanchester combat equations, started. Our work relies on the considerations introduced by Deitchman (1962) and mentioned before. We emphasized that in extreme circumstances a governmental or occupying force might resort to widespread massacres and genocide. Such behavior aims either at intimidating and scaring the population close to the rebels and eventually when the genocide stage is reached to diminish the size of the group who might join guerrilla forces. What might trigger such extreme

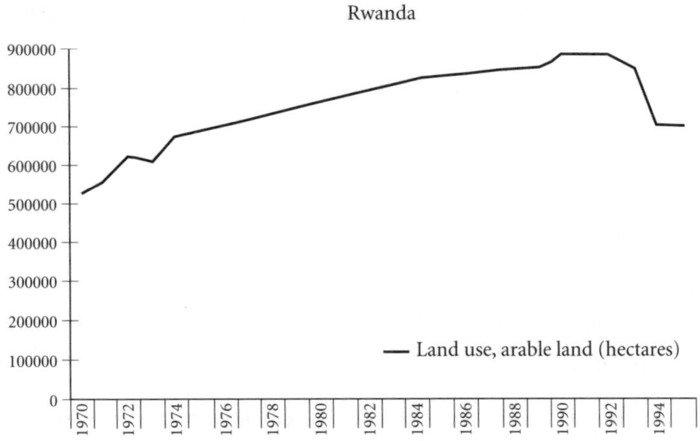

Figure 2 Total Arable Land Surface in Rwanda (source: World Bank).

actions? In our view it was essentially the fear that otherwise rebel groups would even get stronger and take power. We can thus establish the following assumptions for our combat and "Genocide" scenario:

1. The Rwanda situation can be described as a typical Deitchman guerrilla combat model where Tutsi rebels are dispersed but fight the government troops in a concentrated fashion through ambushes. They recruit from about 10% of the total Rwandan Tutsi population (estimated at about 650,000 in 1990 as opposed to 6,800,000 Hutus). Their initial size is estimated from various sources especially Jermann, Sanglard, and Weber in König, Schössler and Stahel (1999) at 5000 in the beginning of 1990. Government troops (mostly Hutus) are estimated at 40,000 and recruitment possibilities for them at about 100 men per week. Tutsi rebels can inflict much heavier losses on government troops than these can on them.

2. The following scenario may be envisaged from 1990 on, consistent with our earlier narratives: The resource crisis due to the overall population expansion leads the (Hutu based) government of President Juvénal Habyarimana to put more pressure on Tutsi controlled land. This leads to an increase in recruits for the Tutsi rebel army which grows rapidly in size. Given the heavy losses this force can inflict upon government troops, parity with the Hutu forces is reached at the end of 1992 and Tutsi fighters continue to deplete them and achieve superiority. Maximum superiority is achieved for Tutsi forces in the spring of 1994. This can be considered in a way as a triggering event for the genocide of the Tutsis and moderate Hutus, which begins in April 1994. In other words, it is assumed here that what triggers the genocide is a desperate attempt on the part of government forces to reduce their differential with the Tutsi fighters. In that sense, the bombing of the Rwanda's President Habyarimana's plane on the 6th of April, the apparent triggering event manifested (whether it was due to Tutsis or extremist Hutus is still unclear) the weakness and loss of control at the top. This then called in the eyes of Hutu extremist and government forces for drastic action to reduce the recruitment base of the Tutsi fighters.

Based upon these assumptions, the following Lanchester type relations can be set up[11]:

$$\frac{dtutsif}{dt} = par1\, pottr - par2\, gov\ tutsif$$

$$\frac{dgov}{dt} = -par3\ tutsif + par4$$

$$\frac{dpott}{dt} = par5\ pott - par6\ par7\ gov\ pott$$

$$pottr = 0.1\, pott$$

$$par6 = 1\ if\ (par8\, gov - tutsif) < 0$$

$$0\ otherwise$$

where *tutsif* stand for Tutsi fighters, *gov* for government forces, *pott* for Tutsi population, *pottr* for recruitment base from Tutsi population. *par1....par8* represent various constant parameters. Three of these deserve further explanation, *par4* represents the drafting of 100 people per week by the government army which was initially trained and supplied by French forces present in the country. *par5* is the rate of increase of the Rwandan population which can be calculated from demographic data up to1994. *par6* represents a logical (Boolean) variable with value 1 when the critical differential mentioned above in point 2, between government forces and Tutsi fighters is reached and 0 otherwise. This critical value has been estimated on the basis of the ratio of forces that drove eventually Tutsi forces to their final victory i.e. at the point when Tutsi fighters are equivalent in numbers to 2.5 government forces. *par6* represents in some sense the "Genocide" parameter[12].

The above differential equations constitute a "typical" Deitchman asymmetric form of the original Lanchester equations with reinforcements where the guerilla (Tutsi) fighters are attacked by government troops in a dispersed way whereas Tutsi forces fight in a concentrated fashion. This relatively simple model gives then the following results expressed in graphical form below (Figure 3).

It has to be pointed out here that reliable combat data for Rwanda are extremely difficult to get. In particular, a monthly evolution of the number of fighters is practically impossible to evaluate. Nevertheless, the swiftness of the Tutsi rebel response after the start of the genocide suggest a relatively effective and superior military force to which allies from Uganda, Burundi and the Congo might have contributed. This conclusion derives from our model and is represented in the Graph of Figure 4. One should also notice that the 2.5 superiority of the Tutsis which triggers the Genocide is close to a 3 to 1 ratio which traditional analyst link to a victorious outcome for the force that achieves it. Despite the Genocide (and maybe because of it) Tutsi superiority is still there at the end of 1994 explaining ultimate Tutsi victory and conquest of power. Moreover the figures we arrive at for the Tutsi rebels (called Rwandan Patriotic Forces) by the beginning of 1994 are a little bit inferior to the post-civil war Rwandan Army mostly now populated by the former Tutsi insurgent forces (around 49,000).

11 This whole system was simulated with the help of the SPARE dynamic simulation package developed at the Graduate Institute of International and Development Studies.

12 Obviously we are saying here that we did not use a statistical procedure to estimate this parameter.

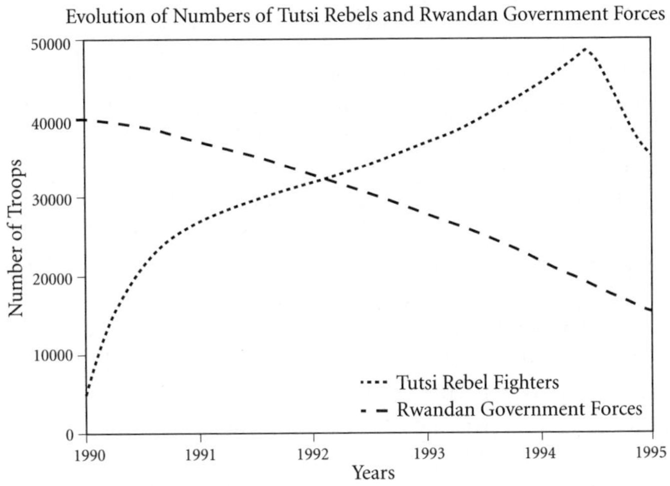

Figure 3 Evolution of Numbers of Tutsi Rebels and Rwandan Government Forces.

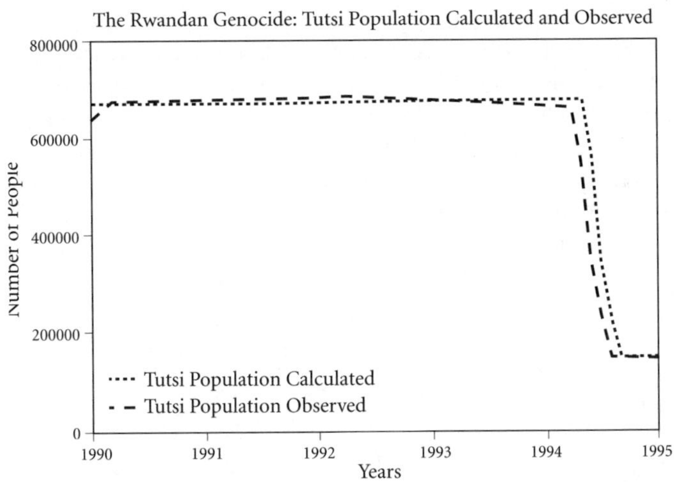

Figure 4 The Rwandan Genocide: Tutsi Population Calculated and Observed – source of real data for population (Kuperman, 2001).

Some more reliable data exists only for the pace of the Genocide and its final magnitude of about 500,000 people. These data can be found in the work of Kuperman (Kuperman, 2001) from the Brookings Institution. The Graph in Figure 4 represents what we can reproduce here solely with the help of our model and without any ad hoc assumption based upon exogenous factors. However more empirical investigations will have to be carried out as additional data becomes available. Since our parameters are not estimated statistically but based upon a priori theoretical considerations, it makes sense here to just test the significance of the correlation between observed and calculated values, which is here 0.98029 and significant at below 0.0001 (t= 37.461, DF=57).

Conclusion

We have tried here to shed some light on the complex linkages between environmental conditions and trends and the issue of conflict as well as the major role of a crucial intervening variable, fear. Through an analysis of some basic aspects of the relationships between natural resources, demography and institutions and an analysis of the existing literature, the Homer-Dixon hypothesis of a direct causal linkage between environmental scarcities and conflict was rejected. However, the importance of essential institutional settings was emphasized. Because the main problem in the end is not environmental but institutional, mostly institution building and consolidating strategies and policies should have the biggest effect in the avoidance of outcomes where environmental scarcities together with demographic expansion and crowding and then fear lead to violence and war lord dominated societies. The biggest task however is to maintain cooperation and prevent conflict in societies most vulnerable to change. Indeed the magnitude and amplification of fighting that leads to massacres and that stems from fear of the other can only be overcome by strong institutions that can guarantee due process of justice within a society. Fear that one party is ready to take over and to initiate revenge only leads to more uninhibited violence[13].

Above all, it is necessary in the future to identify, so that they may be avoided, the positive feedback mechanisms triggered by environmental scarcities that can come about either from increase in population or overuse of resources. These are the mechanisms that can lead to major conflicts within and between societies and that should be curtailed.

This example shows the potential of simulation and modeling in complex situations. In order to investigate cause and effect, researchers must often use a reductionist approach to check the impact of one variable on another. A simulation model of this type, through an iterative process as well as by the inclusion of many different research results, will give a better rendering of the linked effects and hopefully their impact on evolving situations.

References

(2005) "Paradigm in Distress? Primary Commodities and Civil War, Special Issue", *Journal of Conflict Resolution* 49(4).

Allais, M. (1953) "Le comportement de l'homme rationnel devant le risque: Critique des postulats et axioms de l'école américaine", *Econometrica*, 21, 4, p. 503-546.

André. C. and Platteau J.-P. (1998) "Land Relations under Unbearable Stress: Rwanda Caught in the Malthusian Trap" *Journal of Economic Behavior and Organization*, 34, p. 1-47.

Bhavnani, R. and Backer, D. *Localized Ethnic Conflict and Genocide: Accounting for Differences*, Manuscript, Department of Political Science, University of Michigan, Ann Arbor MI.

Brander, J. A. and Taylor, M. S. (1998) "The Simple Economics of Easter Island: A Ricardo-Malthus Model of Renewable Resource Use", *The American Economic Review*, 88, 1, p. 119-138.

13 One can only recall here the terrible words of the King of France, Charles IX when he finally authorized the massacre of the French Protestants, the Saint Barthélémy of August 24, 1572: Qu'on les tue tous, Lets kill them all! As in Rwanda, the massacre did not resolve anything and France would witness twenty years more violent conflict as a result.

CHICHILNISKY, G. (2006) *The Topology of Fear*, Manuscript, Columbia Consortium for Risk Management, Columbia University.
COLLIER, P. and HOEFFLER, A. (2000) "Greed and Grievance in Civil War", in, *World Bank Working Paper Series*, BANK W., Ed., Washington, D.C., World Bank.
DASGUPTA, P. S (1995) "The Population Problem: Theory and Evidence", *Journal of Economic Literature*, 33, p. 1879-1902.
DASGUPTA, P. S. and HEAL, G. M. (1979) *Economic Theory and Exhaustible Resources*, Cambridge, Cambridge University Press.
DEITCHMAN, S. J. (1962) "A Lanchester Model of Guerrilla Warfare", *Operations Research* 10, 6, p. 818-827.
DEMSETZ, H. (1967) "Toward a Theory of Property Rights", *American Economic Review* 57(2), p. 347-359.
DIAMOND, J. (1997) *Guns, Germs, and Steel*, New York, Random House.
DIAMOND, J. (2004) *Collapse: How Societies choose to Fail or Succeed*, Penguin Books.
FEARON, J. and LAITIN, D. (1996) "Explaining Interethnic Cooperation", *American Political Science Review* 90(4) [December], p. 715-735.
FRIEDMAN, M. and SAVAGE, L. J. (1948), "The Utility Analysis of Choices involving Risk", *Journal of Political Economy* 56, p. 279-304.
GAKUSI, A.-E. and MOUZER, F. (2003) *De la Révolution Rwandaise à la Contre-Révolution*, Paris, L'Harmattan.
GLEDITSCH, N. P. and URDAL, H. (2002) "Ecoviolence? Links Between Population-Growth, Environmental Scarcity and Violent Conflict in Thomas Homer-Dixon's Work", *Journal of International Affairs* 56(1), p. 283-302.
HARDIN, G. (1968) "The Tragedy of the Commons" *Science*, 162, p. 1243-1248.
HOMER-DIXON, T. (1994) "Environmental Scarcities and Violent Conflict: Evidence from Cases", *International Security* 19(1), p. 5-40.
JERMANN, P., SANGLARD, H. and WEBER, B., (1999) "Simulating Future Wars", in KÖNIG, E. F., SCHÖSSLER, D. and STAHEL A. Eds. *Konflikte und Kriege, Simulationstechnik und Spieltheorie*, Zürich, VDF Hochschulverlag ETH, p. 115-138.
KAHNEMAN, D. and TVERSKY, A. (1979) "Prospect Theory: An Analysis of Decisions under Risk", *Econometrica* 47, p. 263-291.
KOSTITZIN, V. A. (1937) *Biologie mathématique*, Paris, Armand Colin.
KUPERMAN, A. J. (2001) *The Limits of Humanitarian Intervention: Genocide in Rwanda*, Washington, DC, Brookings Institution.
LANCHESTER, F. W. (1916) *Aircraft in Warfare, The Dawn of the Fourth Arm*, London, Constable.
LEFEBVRE, G. (1973) *The Great Fear of 1789: Rural Panic in Revolutionary France*, New York, Pantheon Books.
LOHMANN, S. (1993) "A Signaling Model of Informative and Manipulative Political Action", *American Political Science Review* 87(2) [June], p. 319-333.
LOTKA, A. J. (1925) *Elements of Physical Biology*, Baltimore, Williams.
PRUNIER, G. (1995) *The Rwanda Crisis, 1959-1994*, London, Hurst and Co.
VOLTERRA, V. (1931) *Leçons sur la théorie mathématique de la lutte pour la vie*, Paris: Gauthier-Villars.
VON NEUMANN, J. and MORGENSTERN, O. (1947) *Game Theory and Economic Behavior*, Princeton, N.J., Princeton University Press, Second Edition.

2.3 Academic Prejudice and the Spirit of Humbleness

by Edmund Harriss[1]

At the 2008 WKD conference the ivory tower I did not realise I was in was rudely shaken. I had not given the cultural divide much thought, following my own path through university and into mathematics. My love of mathematics comes from its power and the way an elegant argument can make the complex simple. My research looks at tilings and patterns, trying to find the simple rules that can generate complex and beautiful images. I also work to try to communicate the beauty and mystery of mathematics in various activites; websites[2], blogs[3], public science exhibitions[4], as well as developing teaching tools.

At the WKD conference I felt a fierceness of rivalry and criticism between different academic disciplines that shocked me. I felt the criticism too often became personal. Much of the rivalry I perceived simply missed the point of what people on the other side were trying to achieve. The fact that it could take place at a conference like the WKD, specifically aimed at communication, scared me. I had gone there expecting to find kindred spirits, like-minded people who shared my own conception of what knowledge embodies; people who would assume that the other participants had something worthwhile to say. Instead, I struggled to suppress frustration at the depths of the differences. To my even greater shock, it gradually dawned on me that other participants might find it just as frustrating to communicate with me.

I wanted to respond directly to this. So, confident in the power of my mathematically trained thought, after the conference I started to write this essay. I was quickly brought back to earth. A social scientist with whom I have had several exchanges since WKD

---

1 Edmund Harriss is currently a visiting professor in the Department of Mathematical Sciences, University of Arkansas.

2 http://www.mathematicians.org.uk/eoh/

3 http://maxwelldemon.com/

4 http://www.tilings.org.uk/shapes

said frankly: "I think you could compare it to me writing an essay on geometry without mentioning Euclide, Thalès and so on, and just talking about the way I feel when I look at triangles, saying that there are big and small triangles, etc."

He had a point. In many ways I was trying to use the wrong tools for the job. I was taking an immensely complex and subtle issue and trying to deal with it in the same way I would build a mathematical proof. We do need to analyse and understand how science works, yet like many scientists I did not have all the skills necessary for the job. Science needs the input of the human and social sciences not least to bridge the gap with the population at large and move scientific understanding into general knowledge. This is not a one way process however. Just as scientists often have problems communicating their understanding, those who study and do research in the humanities can have problems communicating their understanding of science to people working in the physical sciences. This essay is therefore both a plea to others and an exercise for myself; lets take a step back, be humble and try to understand our own ignorances and respect the knowledge of others. It sounds so obvious, yet it is less simple than one might assume. To this end I present my personal view of science and hope that, at the very least, it gives a data point.

Some stereotypes

I define myself as both a mathematician and an artist, so I have always liked to believe that I stand above the great arts versus science debate. However I have to admit that my prejudices and jokes tend to lie more on the science side. Especially as a mathematician I enjoy a cartoon like this:

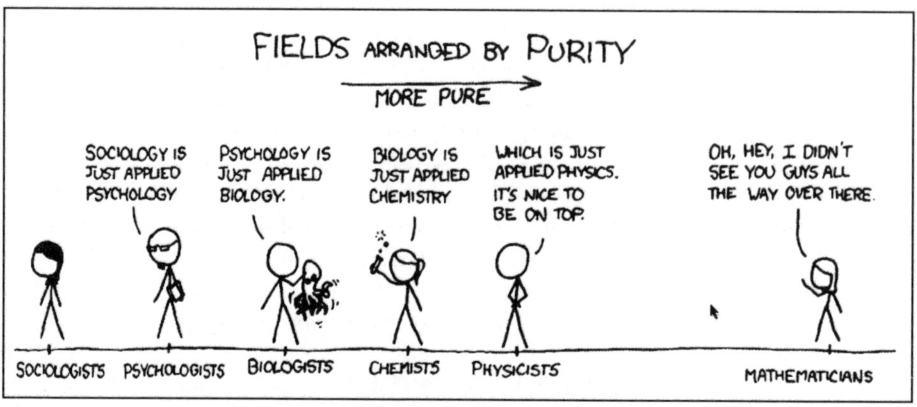

Purity, © Randall Munroe xkcd.com/435.

However I also see there could be alternatives to this view and hope that the account below, though personal, is reasonably balanced.

The cartoon above illustrates a common cliché of the spectrum of knowledge. On the one side lie the disciplines that most reduce the nature of the world to a language of formulae and equations. Chemistry can then be derived from Physics and Biology from Chemistry. From Biology we can develop Anthropology and Psychology, and thus to

Sociology. At the other end of this reductionist representation of the spectrum of human endeavour we might agree to place the Creative Arts. Irrespective of the subjective choice of specific areas that we would place on our personal version of this cartoon, we can probably agree that it resembles a continuum from real life; in every discipline there are people who complain about those in other disciplines in two ways. On one side they say that the results are trivial or that the soul is being removed to reduce everything to an equation[5]. On the other they say the results are not rigorous, or even worse, just empty rhetoric.

At every level, however, there is a remarkable consistency in what people are trying to achieve. They are all trying to understand the world, seeking to find truth, beauty and give meaning both to their own lives and our perception of the world. This is one of the great quests of the human spirit. Whatever we choose to believe in personally, whatever tools we believe can help question or establish our beliefs, we can surely agree at least on one point; Mankind itself is not all-seeing, all-knowing and infinitely-wise and never can be. Even if our neanderthal emotions and stone-age ethics were ever to catch up with our godlike technological capacities, it would be a myth to presume that we could control every parameter that impinges on our individual future or that of our species. Paradoxically, this is not a logical argument against our trying to expand the breadth and depth of mankind's understanding.

Complex is not the same as random

Everyone involved in research, even just at a personal level, has experienced that solving problems in most cases opens more questions than it answers. The world may be vastly complex, but it is not random. There are some points of simplicity. The beauty of science has been to find those points that are so simple they can be put into a mathematical language. Similarly the lack of mathematical models and repeatable experiments at the other end of the cartoon's intellectual spectrum just reflects that more precise (and thus simple) tools often fail because the questions are more complex.

So the reciprocal criticisms described above are both valid, but miss the important point. The right end (in the cartoon)[6] is simple, because it has to be. The left end is not rigorous because it cannot be. Let us start with an assumption, the principle of intellectual equality, that the smart people studying all disciplines of knowledge are equally smart. If a subject seems to do more or less, it is firstly because of the abilities and limitations of the tools we have available, not the shortcomings of the practitioners. This is not to say

---

5 There are even suggestions that engineers are more likely to become terrorists because "characteristics such as greater intolerance of ambiguity, a belief that society can be made to work like clockwork, and dislike of democratic politics which involves compromise, are more common among engineers." Gambetta and Hertog (2009a,b)

6 I found here that putting something so abstract on a linear scale and talking about it are not as easy as they seem. My initial choice to describe relative positions on this scale (from the cartoon) was to use "above" and "below", but these are hopelessly tinged with value. Even "right" and "left" immediately bring up the language of the political spectrum. Nevertheless I have decided to stick with them and add this disclaimer that they only refer to the cartoon.

that there are practitioners who do not have shortcomings, but they too are spread over all areas of knowledge. Of course, the way financial resources are distributed in education and research is not free of market forces. If we can agree on this point, it stands to reason that at some level, it must inevitably have a dynamic feedback on what research areas attract the brightest minds. Yet it would be ridiculous to suggest there is any simple correlation between genius and either financial gain or an easy career path.

The tools available to science improve the whole time, but science defined as the combination of the experimental method with mathematical models, is not slowly pushing aside human argument and debate. Mathematics creates many of the tools and the models (numbers, differential equations, graphs) that are used by science. We have only a limited number of mathematical tools and have to be able to translate the world into the languages they understand. Mathematics, however tells us more than this. It can reveal the limits of our models and even of human knowledge itself. We should accept that we are stuck with two extremes for our methods of understanding and interpreting the world. In moving forward from present to future there is no single correct path, no predefined recipe for success, even if retrospective analyses allow us to perceive and learn from past mistakes. Each approach to understanding provides insights and the best way forward must surely be to ensure that every possible method of enquiry is fully exploited.

Beyond interdisclipinarity

The appeal to a combined approach goes beyond some common notions of interdisciplinary research. In fact an over-simplistic view of interdisclipinary research (often institutionally pushed by insisting departments work together) probably does more damage than good. We must work out how the skills of finding the simple and of truly understanding the complex can be combined. We can never be sure what new avenues of beautiful simplicity remain to be discovered. Yet we must also guard against false assumptions and bad simplification. We need to take better account of how our knowledge can effect what we seek to study. Finally we should consider science itself, how it works and establishes itself. It is not only the mathematician who often seems to forget that, however objective and independent his science, it remains an enterprise conducted by people. Arguably, his science may aspire to purity but he is certainly not a creature of pure logic. As the cartoon cliché reminds us, all acquired knowledge is by definition handed down through human vessels. If you think this is a trivial point, witness how frequently the notions of dialogue between individuals and of dialogue between disciplines of knowledge are freely mixed, in writing and discussion.

This makes for a messy development of knowledge. Let us embrace that, not just as inevitable, but as something we can take advantage of. Specialism should be maintained, yet contacts and personal links should be encouraged as widely as possible. Allowing the possibility for truly wild connections to be made. The true interdisciplinary breakthroughs did not come because two groups of academics were put in a room and told to come up with something. They grew out of individual realisations that two sets of personal skill and knowledge could be productively combined. To achieve this we need humbleness and open minds. We need to accept that even within our expertise there are problems that might be solved, or at least simplified, by a different approach. Similarly we need to accept that when we venture into another's area of expertise they have a good chance of being right if they do not accept our possible solution.

From seeing to science

The scientific model gives us a very thorough understanding of the things that we can apply it to. The laws of physics can tell us about the history of the universe itself. They have also revealed what we are made of at unimaginably small sizes. Humankind's most amazing as well as most awful technical achievements, from the creative and connective power of the internet to the destructive power of an atomic bomb, could not have come about without this scientific understanding and simplification. Furthermore it is only retrospectively that it becomes easy to work out where the simple models can be applied. For centuries of cathedral builders in Europe "The marvellous construction of vault and buttress…was the result of a series of ad hoc solutions to practical difficulties." (Bernal, 1971). It would have seemed clear to them that the rules defining the size of a buttress could not be related to those that said how many windows could be in a wall. Many other rules of thumb, intuitions and vernacular knowledge would apply to every part of the building. Yet nearly all these ideas were swept aside by the mechanics of Isaac Newton, whose principles underlie our ability to build structures today that dwarf those early cathedrals. A more recent example of a physics/maths-based insight into an age-old societal problem (conflict) suggests there are remarkable universal similarities between the dynamics of present-day wars and insurgencies in different parts of the world. These modern conflicts seem to show a power law distribution with coefficient around 2.5. It has been suggested this quantitative similarity reveals information about the structure of conflict, which could radically change our worldview of the issue (Johnson, 2006). Thus even in one of the most chaotic and complex of human activities, war, we may be able to catch glimpses of simplicity.

These two examples illustrate a key factor in the progress of science. To model something you first have to be able to see and measure it. Newton's laws could not have been discovered before the telescope was invented and it is no coincidence that their discovery came a relatively short time after. However one step did not follow automatically from the other; Newton's discovery was still the work of genius.

Today we are becoming able to measure and quantify more and more things. The internet itself offers the possibility of accumulating massive data sets that might reveal hidden structure in many chaotic systems. Genomic organisation and DNA sequencing data offers us access to an incredible wealth of information about nature and humanity. One specific example of where it has revolutionised our knowledge of the past is the study of human population flow. Previously this could only be studied by painstaking archeological methods yielding tiny amounts of data. Being able to read the genetic code of the people now living has opened a whole new window on the past (Cavalli-Sforza *et al.*, 1994). The potential for human genius to exploit such tools will depend on how rapidly and fully it can move beyond outmoded and inappropriate concepts of ownership over fields and techniques.

New models

Eugene Wigner famously commented on the "Unreasonable effectiveness of mathematics" (Wigner, 1960), that it is amazing where mathematics turns up and the links it reveals. I take a slightly different view. Mathematics provides us with certain models of the world; but it is the power of human imagination that has discovered how these models could be applied to the world. Mathematics provides not just some of our models to understand the world, but every model[7] we have. Luckily the study of mathematics is not static and over time new models are created. It is a tribute to human ingenuity that these models often quickly find unexpected or unpredicted use in modelling the world, a famous example being the use of non-euclidean geometry in the general theory of relativity (Gray, 1989). In fact sometimes discoveries already made cannot be articulated as such until the mathematical language has been prepared. In 1931 three respected crystallographers published a detailed study that questioned one of the fundamental assumptions of their field (Goldschmidt *et al.*, 1931). Their work was considered mathematically impossible, so it was not until the 1980s that the structures they had considered were recognised. What had happened? In the meantime mathematicians had discovered aperiodic tilings, such as the Penrose tiling. Such patterns are defined by a lack of translational symmetry. They are highly organized structures yet they not periodic (i.e. there is no patch of tiles that is simply repeated). The strange quasicrystals found by Shectmann in 1984 (Shechtman *et al.*, 1984) could therefore be modelled and thus were able to enter the canon of science (Chapuis, 2003).

The limits of science

As time progresses, therefore, better methods of measurement and new mathematical models combine in unpredictable ways with human ingenuity to open up new areas in which science can be used. However the idea that science will slowly grow to take over all of human knowledge is something we should neither hope for nor fear. It is simply not possible. Firstly, though we will no doubt continue to find surprising ways to measure what could not be previously measured (such as DNA) there will always be many things that can never be measured in any realistic way. As well as constructing the models that mimic selected facets of reality, mathematics can study their limits. One problem that turns up again and again is instability or chaos. Once any system gets sufficiently complicated (and that level of complexity is not high at all) it starts to become unpredictable. Think of throwing a stone with a catapult. The force the catapult uses to project the stone is reasonably consistent. Knowing a few parameters beforehand we can do some calculations and generate a theoretical prediction of where the stone will land. This system is not chaotic. If we fire the catapult several times without changing anything the stones will fall quite close to each other and the place predicted by our model. Now consider the weather. We actually have very highly-developed mathematical models of the weather. Yet, unlike

[7] This is probably a slight overstatement depending on your definition of a model. However anything more than the most basic models have a significant abstract, and thus mathematical component.

our catapult and stone example, if we change the starting data in the model very slightly, the predictions can change dramatically. It is important to note that complex systems are not completely random. In many (such as the weather) it is thus possible to make long-term predictions of average, but not detailed, behaviour. Such sensitivity to initial conditions is called chaotic (Grebogi and Yorke, 1997; Stewart, 2002). Not to forget that many systems we try to study in a quantitative manner (for instance, the epigenetic regulation of our chromatin or the fluctuations of financial currency markets) are far more complex than the weather.

Mathematics has shown there are even stronger limits to our ability to interpret the world through a purely rational lense. The mathematicians Kurt Gödel and Alan Turing achieved something that no scientific method had done before (and few have done since). In adding to the body of human knowledge, they managed to demonstrate it has some inherent boundaries. The two concepts that they discovered are closely related. Gödel showed that any axiomatic system that was powerful enough to include arithmetic could neither be complete or proved to be consistent; A system is considered complete if any statement that can be written in the language is either true or false; A system is considered consistent if it does not claim that both a statement and its negation are true, in other words no contradiction exists. Gödel's theory was developed from a rigorous study of statements of the form "This sentence is false.", which are true if false and false if true. Turing's work is closely related, but developed for computer science rather than logic. He showed that there were some questions, (called undecidable), that could not be answered by any computer program. These ideas are very deep, and have been brilliantly expounded by Douglas Hofstadter (Hofstadter, 1979), so I will not try to explain them further here. Essentially they lead to the conclusion that there are limits to rational or computational research. Initially it was hoped by mathematicians that the limitations would only effect the esoteric corners of their subject. Today, however, we know they touch on many areas, for example the work of Greg Chaitin has shown that undecidable problems run to the heart of number theory (Chaitin, 1998).

Fooling ourselves

Science then is able to find beautiful simplicities and ways of looking at the world and does it in ways that can surprise us. On the other hand as individuals and as a group, our species is faced with many topics that appear urgent to understand and that science cannot tackle now (even if it might in the future). These questions might be about society, education, war or peace, not to mention love and beauty. People who do research in these complex areas should be ready to accept that so-called pure science might still have something useful to say about what they do. In turn, however, scientists working in the simpler areas should be ready to accept that they have something to learn from those who work every day in the more complex settings. This is where communication can make a real difference, no matter how frustrating it may feel sometimes. In particular it is important to realise when the simplifications we have established are false. As an example, take the issue of homosexuality in animals. For a long time it was claimed that there was no data for the existence of homosexuality other than in humans. Yet one of the problems with collecting the data came from species where the males and females are hard to tell apart. A biologist observing the animals could not sex each individual. They would therefore assume any sex act that took place was heterosexual and, where the gender of one animal was known, ascribe the opposite to the other. This particular false simplification was revealednot by a biologist but by someone who trained as a linguist (Bagemihl, 1999). In the quest for simplicity and understanding it is not that unusual for scientists to fool themselves or be fooled. The magician Harry Houdini made a career out of exposing mediums and spiritualists, many of whom had been endorsed by scientists (Houdini, 1924).

Social construcion and logic

This leads to what is perhaps the stickiest point in the clash of intellectual cultures: the debate about whether or not science is socially constructed. I agree with the viewpoint that it is. An alternative version of the debate is about whether science is merely a belief system. Again I agree that at heart it is, but we have to approach both these ideas with subtlety. The scientific method is an effective social construction and provides a very powerful way of producing agreement. This is what allows the common body of knowledge to grow. The strength of science is perhaps better seen not in that it can produce objective truth, but that it can at least identify falsehood. In my experience at least, most scientists spend a long time analysing their methods with care and skepticism. They give deep thought to how an experiment can answer a question and are constantly checking for flaws in their arguments or alternative interpretations of their results. This procedure applies to the discipline of logic itself. In some versions of relativism or social construction our universally accepted system of logic is introduced as just one of many possible, and equally valid, systems. Paul Boghossian has taken this idea on, arguing that it does not have philosophical backing (Boghossian, 2006). I would like to take a different approach. I would love to hear more speculation about alternative logical systems. In fact mathematicians have thought long and hard about this. At the beginning of the twentieth century it was one of the largest topics of mathematical study. A particular system that was studied in depth was Intuitionism proposed by Brouwer as an alternative to the Formalism of Hilbert (Mancosu, 1998). These were essentially the same systems but the first rejected the idea of proof by contradiction (if I can prove that something is false implies it is true then the statement must be true). It led to proofs in many areas of mathematics being re-derived by alternative methods (for example non-standard analysis (Robinson, 1996)). This debate has essentially concluded, though it was not ended by the discovery of some perfect system but by Gödel's arguments that such a system cannot exist.

As a result even mathematics requires some elements of belief; statements that must be assumed rather than proved. Such statements are called axioms and much work has been done on the foundations of mathematics looking at what the simplest collection of statements might be. One of the most powerful methods is to build mathematics from set theory. This is based on the basic idea of a set (naively just a collection of things). The most commonly used axioms for set theory are the ZF axioms (named after their creators Ernst Zermelo and Abraham Fraenkel) (Jech, 2003). These axioms are incredibly simple. For example one (of just 10) states that two sets are the same if they contain the same things. Another axiom says that given two sets the collection of things in one or both of them is itself a set (the union). Personally I find these elements are statements that would be as hard to refute as statements about my own existence, but I accept that they are still unproven beliefs. Mathematicians have not come up with any alternative system with the same power as this classical one, but with different conclusions. This is evidence, though not proof, that no such powerful alternative system can exist.

I believe therefore that it is not useful to apply the concept of social construction to logical systems. Social construction is, however, a powerful tool to help understand how science itself works. Within a restricted area of research one can ask how much the acquisition of understanding affects how the system you study works. Personally I would not give much time to a claim that the discovery of Newton's laws changed the way the universe worked, but we have no way to test this. In Biology, however, it is regarded as usual good practice to make many controls designed to minimise (but not eliminate), the risk that an experimenter's actions have inadvertently influenced the outcome of his experiment or his interpretation of data. A wrong decision can lead to bad data, as in the the study of homosexuality in animals mentioned above. If the interpretation of their research is called in to question, many researchers react emotionally as if their personal integrity had been doubted. However it can be illuminating to examine how much supposed facts evolve over time. One comprehensive illustration of this point can be found by comparing modern text books on general human physiology and medicine with some of the reputable texts from one hundred years ago (Porter, 1997).

At the other extreme of the cartoon's scale, there are situations where there is universal agreement that the objects of study are socially constructed; the classic example being money. In the science of economics, there are simple rules to be discovered. The relationship between supply and demand and prices for example can be modelled. In this case the understanding is pushed back into the system. People hoard goods in an attempt to bring up the price and so on. Economic understanding is overtly used within the system that it models so understanding the rules creates new, more complex rules. In such circumstances we should hold a subtle balance. On the one hand we should not reject the power of the scientific method to provide simplification. On the other we should not reject ideas because they are not part of the model. It is in subjects such as this that we need the most collaboration between styles of thinking. Yet, at least to my outsider's perspective, it appears as the area where the most disagreement between academic cultures takes place. As a result, whatever precious understanding we can achieve by any method is often widely ignored and the forces of popular and financial opportunism have a free hand (Weinstein, 2009).

The politics of science

Now consider the realm of natural science itself (or equally the humanities). These are immensely complicated societies that are able to generate new knowledge. Sometimes the process seems to come about more by accident than design, though we may need to believe that we plan it. Anyone who has worked in a science department at a university is conscious of an element of internal politics that goes on not to mention the fashions and mini-trends of funding decisions and the biases introduced by overfunding of certain viewpoints. To illustrate this with a historical example, let me commit heresy for a British mathematician; Liebnitz did better than Newton. We shall leave aside the controversy of who actually came up with the calculus. Liebnitz came up with the far superior dy/dx notation, the that we use today. Newton just put a dot over the function for his fluxions. The political rivalry, however, meant that British mathematicians for years afterwards continued to use the less flexible Newtonian notation. It can be argued that it is not until the work of G H Hardy at the start of the twentieth century that British mathematics caught up to the rest of Europe (Wiener, 1949). This sort of issue will get more problematic and complex as the number of people involved in science continues to grow. There is a wealth of excellent literature discussing the nature of science and this merits much more consideration by scientists than it has received. For example Bruno Latour has analysed the ways in which scientific work becomes fact (Latour, 1987). He considers how papers and scientists draw upon the work of the past to position and defend their work and how this is in turn reinterpreted by later papers. Eventually scientific results become black boxes that can only be opened with effort. An example is the first discovery of quasicrystals discussed earlier, a paper that was initially forgotten, yet has now been given a new lease of life as an example of a scientific chance missed. Another insightful analysis of the nature of science is that of David Turnball, who has proposed that our modern system of knowledge is not so different from ancient ones and that "modernity's drive for order concealed its messy contingent, unplanned and arational character." (Turnbull, 2000).

So let me finish by reiterating the plea of Julia Higgins (Higgins 2008). Let us all be aware of the failings of our own subjects and adhere to the principle of intellectual equality. Let us come to acknowledged experts in other fields with an initial respect; listen to what they have to say about our subject, tell them what we think of theirs and trust that they will listen and not just reject without thought. Finally remember that communication between different levels of the original cartoon's specturm is hard. When we fail it is much more often because of the complexity involved, than because of any stupidity or ignorance.

References

Bagemihl, B. (1999) *Biological exuberance: animal homosexuality and natural diversity*, St. Martin's Press, New York, 1st edition.

Bernal, J. D. (1971) *Science in History*, M.I.T. Press.

Boghossian, P. A. (2006) *Fear of knowledge: against relativism and constructivism*, Clarendon Press, Oxford.

Cavalli-Sforza, L. L., Menozzi, P., and Piazza, A. (1994) *The history and geography of human genes*, Princeton University Press, Princeton, N.J.

Chaitin, G. J. (1998) *The limits of mathematics: a course on information theory and limits of formal reasoning*, Springer.

Chapuis, G. (2003) "Crystallographic excursion in superspace", *Crystal Engineering*, 6(4), p. 187-195.

Darbellay, F., Cockell, M., Billotte, J., and Waldvogel, F., Editors (2008) *A vision of transdisciplinarity: Laying foundations for a world knowledge dialogue*, EPFL press.

Gambetta, D. and Hertog, S. (2009a) "Can university subjects reveal terrorists in the making?", *New Scientist*, (2712).

Gambetta, D. and Hertog, S. (2009b) "Why are there so many engineers among islamic radicals?", *European Journal of Sociology*, 50(02), p. 201-230.

Goldschmidt, V., Palache, C., and Peacock, M. (1931) "Uber calaverit", *Neues Jahrbuch für Mineralogie*, 63, p. 1-58.

Gray, J. (1989) *Ideas of space: Euclidean, non-Euclidean, and relativistic*, Clarendon Press, Oxford, 2nd edition.

GREBOGI, C. and YORKE, J. A. (1997) *The impact of chaos on science and society*, United Nations University Press, Tokyo.

HIGGINS, J. (2008) "The rules of the game for a knowledge dialogue", in Darbellay *et al.* (2008).

HOFSTADTER, D. R. (1979) *Gödel, Escher, Bach: an eternal golden braid*, Basic Books, New York.

HOUDINI, H. (1924) *A magician among the spirits*, Harper brothers, New York and London.

JECH, T. J. (2003) *Set theory*, Springer, Berlin, the 3rd millennium ed., rev. and expanded edition.

JOHNSON, N. F. (2006) *The mother (nature) of all wars? Modern wars, global terrorism, and complexity science*, APS News, 15(10).

LATOUR, B. (1987) *Science in action: how to follow scientists and engineers through society*, Harvard University Press, Cambridge, Mass.

MANCOSU, P. (1998) *From Brouwer to Hilbert: the debate on the foundations of mathematics in the 1920s*, Oxford University Press, New York.

PORTER, R. (1997) *The greatest benefit to mankind: a medical history of humanity*, W. W. Norton, New York, 1st american ed edition.

ROBINSON, A. (1996) *Non-standard analysis*, Princeton University Press, Princeton, N.J., rev. ed edition.

SHECHTMAN, D., BLECH, I., GRATIAS, D., and CAHN, J. W. (1984) "Metallic phase with long-range orientational order and no translational symmetry", *Physical Review Letters*, 53, p. 1951-1953.

STEWART, I. (2002) *Does God play dice?: the new mathematics of chaos*, Blackwell, Malden, Mass., 2nd edition.

TURNBULL, D. (2000) *Masons, tricksters, and cartographers: comparative studies in the sociology of scientific and indeigenous knowledge*, Harwood Academic, Australia.

Weinstein, E. (2009) "A science less dismal: Welcome to the economic manhattan project", in *The Economic Crisis and It's Implications for The Science of Economics*, Perimeter Institute. http://pirsa.org/09050047.

WIENER, N. (1949) "Godfrey Harold Hardy 1877-1947", *Bull. Amer. Math. Soc.*, 55(1), p. 72-77.

WIGNER, E. P. (1960) "The unreasonable effectiveness of mathematics in the natural sciences", *Communications on Pure and Applied Mathematics*, 13(1), p. 1-14.

2.4 Kachile – Concepts, Tools & Strategies for a Post-Conflict Environment

by Ulf Richter[1]

Côte d'Ivoire is a young country with a fast growing population, estimated at 20 million people (IMF, 2009). Located in West Africa and bordering the countries of Liberia, Ghana, Guinea, Burkina Faso and Mali, it is ethnically diverse with about 60 languages spoken (CountryWatch, 2007). The country is the world's largest cocoa exporter; an estimated 6 million people live directly from cocoa (The World Bank & International Finance Corporation, 2008).

Today, Côte d'Ivoire, once the miracle of West Africa, is in a desperate state. After a failed *coup d'Etat* in 2002 that was followed by an internal conflict with an estimated 3000 victims, Côte d'Ivoire is now a post conflict environment with an UN mission of about 9000 soldiers and staff (MINUCI, 2004). The first post-conflict elections have been postponed numerous times and have yet to take place. In addition to conflict related issues, the country is facing massive social and environmental problems: substantial unemployment (often reaching 40-50%), increasing poverty rates, a large informal sector, an alarming HIV/AIDS rate and high Malaria risk, substantial pollution, high corruption, low trust in (partly dysfunctional) governmental institutions, a high rate of immigration, lack of waste management, and deforestation (The World Bank & International Finance Corporation, 2008; Transparency International, 2009; UNAIDS, 2009; WHO/AFRO, 2004). However, there is hope on the horizon: the country is now generally peaceful, has just negotiated a debt relief agreement with the IMF, and has been affected relatively little by the global financial crisis. Foreign investors are slowly coming back to the country and the awareness of the need for change is rising.

In this article, I discuss strategies and tools for a post-conflict environment, outlining the recent developments at Kachile, a social venture operating in Côte d'Ivoire. I start out

1 Ulf Richter is a Professor and Principal Investigator at the Pontifica Universidad Católica del Perú. He has been a visiting research associate at the University of Grand Bassam, Côte d'Ivoire.

with an introduction to Kachile, its goals, projects, strategies, management and leadership strategies. I conclude with some final remarks on the future of Kachile and entrepreneurship in West Africa.

Kachile

Kachile is a social venture dedicated to alleviating poverty and raising environmental consciousness in Africa. Kachile functions as a think tank for sustainability solutions, combining world-wide expertise and local in-depth knowledge. Kachile regards itself as a hub for entrepreneurs and change agents in West Africa that drive social innovation and transformation in fields such as education, health, ecology, environment and enterprise development. The word Kachile means "change" in the local language Baoulé. Kachile aims to achieve gradual and lasting social transformation through a number of projects, including:

1. kachile.com, an e-commerce for artisanal products, initially from post-conflict Côte d'Ivoire and later more broadly from West Africa;
2. establishment of an Information Communication Technology (ICT) & Entrepreneurial Learning Center in Grand Bassam;
3. a local fashion label, drawing on the existing skills and talent in Grand Bassam;
4. the import of products with a high social impact such as used bicycles for Ivorian youth and solar lamps to provide light to village people and reduce the dangers from open fire and kerosene lamps;
4. research on a holistic approach to microfinance, including the use of mobile technology, increasing financial literacy of families and children, entrepreneurship, and management skills.

Kachile will direct ICT dissemination towards the creation of inclusive markets by fostering entrepreneurship, spurring endogenous growth, reversing a psychology of poverty, and making lasting change at the bottom of the pyramid.

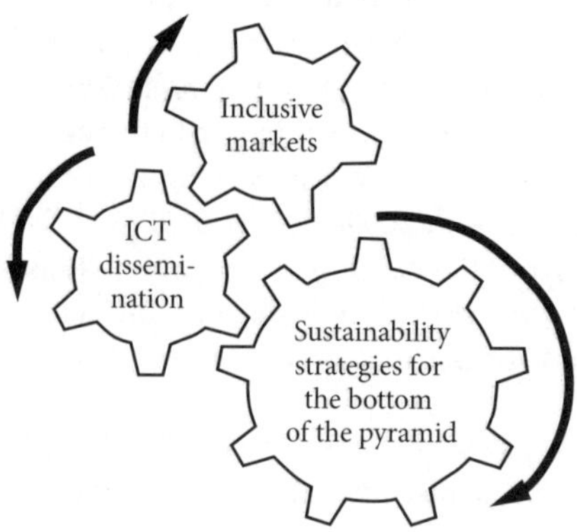

Kachile is headquartered in Grand Bassam, Côte d'Ivoire. Grand Bassam is a small city with about 70,000 inhabitants, covering about 11,300 hectares. It is located about 30

kilometers East of Abidjan, with four million inhabitants the largest city of Côte d'Ivoire and the major port of West Africa. Grand Bassam was the first French colonial capital from 1893 to 1900 and remained a key seaport until the growth of Abidjan beginning in the 1930s. In 2008, the city of Grand Bassam applied for being nominated as a UNESCO world heritage site to protect the unique architecture and landscape of its colonial district – the Quartier France. With its beautiful setting, 15 kilometers of fine sand beach, the surrounding lagoon Ébrié and the romantic colonial district, it remains a key destination for local tourism and an active artisanal center (Moulod *et al.*, 2007; N'Guessan, 2003). However, due to the crisis very few international tourists still find their way to Grand Bassam; most artisans are without work.

E-Commerce for Artisanal Products

Vision and Mission: kachile.com aims to become the n° 1 platform for promoting West African art and selling the products of West African artists and artisans by creating digital opportunity through the targeted application of information and communication technologies (ICTs). It will operate under market dynamics, moving away from pure charity and philanthropy to a pro-poor business model.

Business Opportunity: E-commerce for traditional and contemporary art and artisanal products from West Africa. High-quality artisanal products from West Africa are currently exchanged at enormous price differences between the local selling price of the artist and the final price for the consumer in Europe or the United States. The price of a product may go up to ten times of its original value. At present, such products are mostly brought to the European markets and the US either by individuals, mostly part of the African Diaspora, or by merchants who fill whole containers with products and ship them to Europe or the US to resell them to smaller shops. Development of an e-commerce will allow producers and suppliers operating on the ground to have direct connection with consumers, small shops, art galleries and other potential customers such as banks or corporations. This will allow for higher profit margins for local artists and lower prices for European and US customers.

Products: Initially, kachile.com will market artisanal products from West Africa. At present these are exclusively sourced from the artisanal village of Grand-Bassam, which represents one of the major centers for artisanal products of Côte d'Ivoire. Many of the artists and antiquaries are located in the historic Quartier France – a peninsula in between a lagoon and the Atlantic Ocean which was once the colonial residence of the French. Products include objects of African art such as masks, ceramics, local fashion, woven fabrics, music instruments, figures and statutes, furniture, and jewelry.

Business Model and Strategy: kachile.com aims to generate income through direct sales and online sales of artisanal products. From a commercial perspective, the central issues that kachile.com addresses is the current lack of direct access for local African artists to global markets and the lack of promotion of African culture to mainstream consumers. The approach: Kachile.com will be set up as a data-rich, transparent e-commerce platform providing access to potential customers around the world, using the latest web 2.0 technologies. Kachile will use state-of-the-art software solutions for e-commerce, email marketing, database management, graphics & design, and web-based project management tools to administer its operations. To distinguish itself from existing e-commerce websites for artisanal products, kachile.com will pursue a strategy of providing extensive background information including videos on artists, ethnicities, meanings and traditional use of products, and local African culture.

Target Market and Marketing: The target market consists of a) West African corporations and philanthropists that are interested in promoting local African culture, and b) middle class consumers, wholesalers, art galleries in Europe, Latin America, and the United States that are interested in African, or more generally, ethnic products. Specifically, Kachile aims to organize (i) fashion shows with models wearing African style fashion and jewelry, (ii) African parties where African culture is presented in a private setting organized by individuals with African roots or strong connections, with the ambiance set through African music, movies, and presentations, and where Kachile products are presented by the organizers; (iii) expositions of African art in galleries, restaurants, banks, schools & universities, fashion selling shops and other free or low-cost venues, (iv) product fairs for Kachile products located in city markets, administrative buildings, banks, corporate headquarters, and shopping centers. Finally, (v) kachile.com will heavily rely on online marketing, including direct emailing of its product catalogue, the use of social networking sites such as Twitter, Facebook, LinkedIn or Xing, blogs and Google ads.

One may ask, in what way are such forms of aggressive commercial marketing in any way different from the historical approach of most commercial trading ventures that have emerged in "developed" parts of the world? The essential hallmark of Kachile lies in the intention to develop it along the trade model of a "cooperative society". That is to say, with the primary goal of redistributing the income generated, throughout the strata of local society. Imperative to our model is the notion that recent advances in social networking media serve not merely as one-way conduits for marketing strategies. Such networks need to be respected as vehicles for bidirectional exchange that adheres over the long term to the principles of equitability and reciprocity.

Social Impact: Kachile aims to create substantial social impact through its operations. There are about 1,000 artists working only in Grand Bassam. Considering the typical family sizes in Côte d'Ivoire, there are about to 5,000 to 10,000 to people directly depending on their work. Lifting their living standards will be a major immediate goal. Kachile aims to replicate its model at a later stage all over West Africa.

State of operations: Kachile is currently working on mapping the artisanal landscape in West Africa, developing its website www.kachile.com, and producing a first product catalogue. The organization has received first orders from wholesalers in the United States, France and Austria, and entered into negotiations with logistics and shipping handlers to receive favorable shipping terms. In addition, it has initiated the administrative process of application for a fair trade label that is acknowledged in the global marketplace.

ICT & Entrepreneurial Learning Centers

A focal element of Kachile's philosophy lies in its multi-stage approach to entrepreneurship and empowerment by fostering lasting and sustainable business solutions through knowledge transfer in ICT & Entrepreneurial Learning Centers. This will be achieved through three major axes:

(i) setting up of IT infrastructure, technological support and necessary ICT instruction;
(ii) business coaching;
(iii) incubation for local SMEs.

Concept: Kachile aims to direct ICT dissemination and microenterprise creation towards social and environmental goals, including, but not limited to, creating employment for young adults, in particular women, improving the access to essential medicines, and fostering entrepreneurial ventures that focus on waste management and renewable energies. The objective is to (i) create jobs and bring marketable business skills and build professionalism to Ivorian youth; (ii) offer vocational and executive tools supporting poverty alleviation and post-conflict recovery; and (iii) mobilize ICTs to serve African people.

Implementation: Kachile plans to initially set up well equipped ICT & Entrepreneurial Learning Centers that will serve for ICT instruction and as cyber cafes in major locations in West Africa. The centers will be run as for profit-business during part of the time and used as class rooms for teaching computer and business skills during the remainder. We envision different difficulty levels of courses taught to adapt to different age groups, levels of education, business training and individual needs. An important aspect will be to train local youth to enable them at a later stage to teach courses to their fellows.

Course Content: The course content will be based on existing course materials from courses taught by myself and colleagues from academic institutions around the world. New courses will be developed in an adaptive fashion in response to local needs and based on an impact assessment of past courses. Kachile hopes to further draw on existing resources provided by partnering educational institutions in the future.

Pilot Project: A pilot is planned in collaboration with local high schools in Grand Bassam. Later, when further up the learning curve, the same concept will be replicated throughout the whole of Côte d'Ivoire and West Africa. The recently opened export processing zone and high tech park VITIB already provides state-of-the-art facilities equipped with modern information technologies and represents and alternative location for the pilot project. Kachile will later expand its full range of activities to the rest of Côte d'Ivoire, in particular to the areas that have been affected by conflict. Kachile plans to serve as an incubator for small businesses that will emerge through its entrepreneurship courses and venture plan competitions. Seed funding (derived from the income generated by its commercial operations) will help nascent entrepreneurs to establish their ventures. Kachile will acquire equity in the spin-off businesses generated out of its activities and this is expected to become its main source of income over the midterm.

Target Group for entrepreneurial coaching: Kachile will mainly target the young adult population in the age range 18-30 years, being conscious of the requirements of the post-conflict environment of Côte d'Ivoire. Not only is unemployment above average in this age group but also its adaptability to new information and communication technologies is assumed to be particularly high. Kachile also aims to be able to provide new skills to ex-combatants who are looking for ways to reintegrate into the Ivorian society. Special courses will be provided foster an enabling learning environment for female entrepreneurs.

Income generation: Experience shows that paid services have a better reputation than free offers. Even low-income groups will be charged a minimal fee for access to entrepreneurial coaching courses and for internet access. ICT & Entrepreneurial Learning Centers will aim to generate income enabling them to gradually expand their operations through (i) the revenues generated by their activities (ii) seeking donations and grants.

Fashion Brand

Kachile's initiators believe that branding Africa and African products as fun, fresh, and different is a key factor to success for sustainable business solutions and to overcoming the psychology of poverty in post-conflict Côte d'Ivoire. For too long Africa has been seen as a dark continent of disease, poverty, pollution, and war. The time has come to instill a new image of Africa as a place of art, craft, fashion, culture and design.

Concept: The group plans to develop a West African brand for fashion and artisanal products, to be marketed and distributed through the e-commerce website kachile.com. Our goals are (1) to brand African products as desirable and fun, (2) to create local economic opportunities, (3) promote female empowerment, and (4) promote ICT dissemination. More specifically, the ideas are to design for existing and future products suitable for both African and Western consumers, train locals (in particular women) in the use of software & internet, establish the necessary logistics, market the branded products, and deliver them to the consumer.

Impact measurement: With this project Kachile aims to create direct impact by providing economic opportunities to local people – one person employed generally supports 5-10 family members. Success will be measured by quantitative monitoring; of products being sold, overall number of people employed, proportion of women employed, and brand awareness.

Import of Affordable High Impact Products

A key problem for the local population is the general lack of affordable products that respond to essential needs such as health, transport, or illumination at night. Kachile is currently negotiating with a number of providers to import and distribute products that respond to such needs. This includes:

Used bicycles: Kachile is likely to receive financing from a social investor from Switzerland to start importing used bicycles from Switzerland in the summer of 2010. The bicycles are indirectly subsidized by the Swiss government as part of a larger social project to let unemployed persons in Switzerland recycle old bicycles for sale to developing countries at a reduced price.

Solar powered lamp: D.Lights[2], an international consumer products company, designed the award-winning Kiran solar lamp which costs about US $10 per unit to pro-

[2] http://www.dlightdesign.com

duce. An estimated 1.5 billion people in the world live without access to electricity. Many households rely on kerosene lanterns and candles to meet their lighting needs, but these light sources are expensive, hazardous, polluting and produce low-quality light. By replacing a kerosene lamp with a D.Light lamp, a consumer can expect to experience cost savings of up to US $150 over five years; increased safety due to the elimination of accidental fires caused by kerosene lamps; better health due to the reduction of indoor air pollution; and increased productivity as household and livelihood enhancing activities can be performed at night. A second organization, SunNight Solar[3], sells similar products and is equally interested in expanding into Côte d'Ivoire.

Affordable eyeglasses: Another product idea focuses on enabling poor people in developing countries to buy affordable eyeglasses. 400 million people in the developing world need eyeglasses to do their work; in developing countries, eyeglasses typically cost between US $40-60, a price truly out of reach for people living on less than US $4 a day. Without eyeglasses, workers lose the ability to earn income, forcing their families and communities into deeper poverty. However it is logistically challenging to distribute eyeglasses on a large scale. In order to reach millions of people with a program that will eventually become self-sustaining, an entrepreneurial business model is needed. VisionSpring[4] has devised a project that provides affordable eyeglasses on a large scale with a low cost structure by empowering local entrepreneurial women. VisionSpring sold 400,000 eyeglasses in 2009 at an average cost of US $4.

Healthcare: Medication is very expensive in Côte d'Ivoire. People can easily die simply because they cannot afford essential drugs like Paracetamol. Life expectancy is below 50 years. This is not primarily because of AIDS/HIV as many people believe. Malaria and poverty-related issues are a much bigger problem. However, the AIDS rate in particular in the urban slums of Abidjan is very high (up to 25% of young females). Kachile is currently exploring potential import of affordable generic drugs from countries such as India and how to manage the distribution.

Current Partners

International University of Grand Bassam: The International University of Grand Bassam (IUGB) is envisioned as a Regional Center of Excellence that will help educate and train the human capital required for sustained growth and development in West Africa. IUGB opened in January 2005 and was formally established as an accredited institution of higher education by Presidential Decree 2007-477 of May 16, 2007. IUGB has established an active partnership with Georgia State University. IUGB students now take two years of course work at IUGB and then transfer to GSU to complete their majors. They currently receive their degree accreditation from GSU. As of the fall semester beginning in September 2009, IUGB employs 29 part time and 14 full time faculty, among them 19 hold a PhD or doctorate from US, Canadian and European universities. The number of students has grown from 8 students in January 2005 to 136 in the fall semester 2009. IUGB offers

3 http://www.bogolight.com
4 http://www.visionspring.org

thus an outstanding teacher to student ratio. 24 IUGB students are currently studying at Georgia State. IUGB celebrated the graduation of its first student from GSU in June 2009. IUGB is currently creating a Center for Innovation and Entrepreneurship (CINOVE) to trigger entrepreneurial activity in Africa by offering know how, tools, business plan competitions and networking activities for active and nascent entrepreneurs. Special attention will be paid to the potential of information and communication technologies, biotechnology, agribusiness, and renewable energies. CINOVE will be specifically tailored to the African business environment, taking into account local needs and challenges such as the importance of SMEs and the informal sector.

Fondation TAPA: Fondation TAPA is a non-profit organization headquartered in Abidjan, Côte d'Ivoire, that aims for the promotion of African culture. TAPA was founded in August 2001 and strives to develop the African museum institutional capacities, promote contemporaneous art in Côte d'Ivoire, and build a bridge between Africa and other continents. Its objectives include promoting Ivorian artists and products throughout to the world, providing state-of-the-art information on the art sector, organizing events for artists and art professionals, and intensifying Ivorian cultural practices. Originally from Côte-d'Ivoire, "tapa" refers to a material made with beaten wood bark used as traditional clothing. Fondation TAPA has chosen its name to portray its commitment to rehabilitate, promote and confer a title of nobility to the African arts sector.

OPDENTCI: OPDENTCI is a private export promotion and marketing organization for artisanal products from Côte d'Ivoire. OPDENTCI aims to provide stable and sustainable incomes to artisans that lack business skills to market their products. OPDENTCI represents 34 cooperatives from Côte d'Ivoire, potentially benefiting about 30,000 artisans in Côte d'Ivoire and the ECOWAS region. OPDENTCI was invited to participate in the general assembly of the UN Global Compact Participant in November 2007 and has since then made the principles of the UN Global Compact part of its day to day operational actions and strategies.

Strategy & Management

Kachile has a number of advantages over existing commercial operations and cooperative societies based in West Africa, benefiting from its international network and its unique positioning in Côte d'Ivoire.

Team: At present, Kachile is run entirely by volunteers. Kachile is expecting about 20 volunteers from Europe in Grand Bassam in 2010 to assist its operations and to help in implementing the set up of e-commerce and the ICT Learning Centers. Kachile is also supported by an international network of business professionals and academic experts and collaborates with legal experts, designers and IT professionals from Côte d'Ivoire. Kachile is advised by internationally renowned strategy and development experts and is currently negotiating with a number of industry experts and public figures to become part of the advisory board to help develop and promote Kachile.

Local embeddedness: Kachile will focus on responding to the needs of the informal sector, private companies and the public sector of the West African region. Kachile works with local cooperatives, the TAPA Foundation, representing Ivorian artists, OPDENTCI, representing 22 Ivorian cooperatives of artisans, and with a local team that helps in sourcing of quality products. As a Research Associate and former Assistant Professor at the International University of Grand Bassam I myself am well positioned to access local resources and to gain credibility for Kachile's activities.

Collaboration: Kachile will pursue an elaborated partnership strategy to create collaboration with corporate and academic partners from around the world interested in fostering entrepreneurship in Africa. In addition, Kachile will develop an outreach strategy to achieve dissemination to the wider community. To foster youth engagement, Kachile aims to develop strong links with the student body of the International University of Grand Bassam, and later with other universities throughout West Africa. Kachile is also currently negotiating with the German ministry for development. Kachile emphasizes that while it is aiming to be a change maker, it has very limited financial means. To be successful in the long run Kachile seeks to join up with a number of different partners, among them local cooperatives, universities, development agencies, NGOs, government agencies, UN institutions, media, and corporate partners in West Africa.

Income generation and attribution: Kachile is a mission-based venture. Kachile will generate income through (i) profit-sharing from its e-commerce, (ii) revenues generated by ICT, business and entrepreneurship courses in ICT & Entrepreneurial Learning Centers, (iii) income generated through microfinance and successful investment in SMEs, (iv) donations and grants, and (v) consulting services. Kachile aims to attribute its income to community projects, SME creation, scholarships for ICT, business and entrepreneurship courses, endowment, and research projects.

Legal structure: Kachile was founded in January 2009 and to date is mainly operating in Côte d'Ivoire. The association has adopted a hybrid structure to be able to realize the full range of its projects. It is registered as a Swiss association operating in post-conflict Côte d'Ivoire. For fundraising purposes tax-exempted associations are planned in Germany, France, the UK, and the United States. Kachile.com has been registered as limited company in Ivory Coast to be eligible for export promotion programs.

Conclusion

Côte d'Ivoire needs change if it is ever to re-emulate the position as political, cultural and economic leader in West Africa that it once held. Many institutions are not functioning and it remains unclear when the Ivorian elites will eventually turn their attention back to reconstructing the country instead of fighting for power. The Kachile project is a response to the institutional void, the lack of political will, and the ubiquitous poverty caused by a combination of mismanagement by the local elites and mistreatment by the international community over the last 20 years (Klaas, 2008). Despite the challenges ahead, the response over the year 2009 and early 2010 from the local people in Grand Bassam, the academic community interested in developing countries and poverty alleviation strategies, youth from around the world who want to come help setting up operations, and social investors that are interested in financing Kachile projects, has been amazing. No matter how modest or small, the Kachile project is promising to bring about a change of attitudes that can further procreate, create employment, deepen understanding of development issues, and help in creating new relationships between West African people and the rest of the world.

References

CountryWatch. (2007) *Côte d'Ivoire Review 2007: Chapter 1: Country Overview*: CountryWatch.

IMF. (2009) *IMF Executive Board Approves US$565.7 Million PRGF Arrangement for Côte d'Ivoire, Press Release No. 09/96 of March 27, 2009*. Washington, IMF.

Klaas, B. (2008) "From Miracle to Nightmare: An Institutional Analysis of Development Failures in Côte d'Ivoire", *Africa Today*, 55(1), p. 109-126.

Minuci. (2004) Côte d'Ivoire - MINUCI - Facts and figures. http://www.un.org/Depts/dpko/missions/minuci/facts.html. Accessed April 13 2009.

Moulod, J.-M., Kongo, J., and N'Guessan, S. (2007) *Akwaba à Grand Bassam*, Grand Bassam, City of Grand Bassam.

N'Guessan, K. N. D. (2003) *Grand Bassam: Ville culturelle, Ville de patrimoine*. Grand Bassam: City of Bassam.

The World Bank & International Finance Corporation. (2008) *Doing Business 2009: Country Profile Côte d'Ivoire*. Washington, The World Bank & International Finance Corporation.

Transparency International. (2009) *Corruption Perceptions Index 2008*. Berlin: Transparency International.

UNAIDS. (2009) Ivory Coast. http://www.unaids.org/en/CountryResponses/Countries/ivory_coast.asp. Accessed April 13 2009.

WHO/AFRO. (2004) *Malaria Country Profiles: Côte d'Ivoire*, Geneva, WHO/AFRO.

2.5 Disciplines for the Common Good: From insular to systemic interdisciplinarity

by Filippo Dal Fiore[1]

The issue of mutual acknowledgement and respect among different knowledge branches is pivotal when addressing complexity and solving problems in our interconnected world. Any disciplinary solution proposed as "the ultimate blueprint" is likely to have unforeseen repercussions in other realms, the role and importance of which would be underestimated. As a consequence, global decision makers can end up pursuing problematic solutions or even disregard expert knowledge altogether and follow their own agendas.

In this paper, I argue for the necessity of different disciplines converging towards priorities determined by a shared notion of common sense and the common good. In parallel, I justify the need of complementing what I define as *insular interdisciplinarity*, or explorations within a scientific community around a well-defined topic, with a more *systemic* approach that tracks the interdependencies among different macro-topics which are relevant to a well-defined issue.

The challenge ahead of us

The nature of problems in today's world is complex, as they are not self-contained but have co-evolved and are interconnected (Jervis, 1997). This implies that, in order to thoroughly understand them, we need to tackle them as pieces of a larger puzzle as well as observing them in isolation.

1 Filippo Dal Fiore is a researcher at the Senseable City Laboratory of the Massachusetts Institute of Technology, Boston and a co-founder of Currentcity, an R&D foundation dedicated to exploring the applicability of digital data analysis to urban monitoring.

As an example, let's consider the issue of global warming. First, we need to define the problem scientifically, relate it to our existing knowledge in the field of meteorology, and also observe its manifestation through the latest methods and tools of the natural sciences. Subsequently, we need to understand why the problem exists in the first place, hence the need to address topics like climate cycles, economic growth, human ethics, polluting technologies, lifestyles, ecosystems, among others. Ultimately, in order to tackle and potentially solve the problem, we need politics, common intents, financial incentives, negotiation, individual action, behavioral change, sustainable technologies, theory as well as practice, rationality as well as emotions.

Expert knowledge exists within all the areas listed above and certainly still has plenty of room for growth. What most often is missing is their mutual acknowledgement, as well as a sense of partnership between experts to solve the same problem (i.e. climate change as a whole), defined and prioritized in the same way. Each discipline addresses a subsection of it, shedding light from its unique angle, but perhaps also underestimating the way in which the content of the observation is the complex manifestation of many other forces and therefore should not be reduced to any subset of them (Anderson, 1972).

Any disciplinary solution implemented as the ultimate blueprint to address a given problem is likely to have unforeseen repercussions in other realms: for example, ambitious economists may push an economic solution, e.g. a new policy, while underestimating its political, environmental and cultural implications. Ambitious engineers may push a technological solution, e.g. a new technology, while underestimating its economic, cultural and political implications. Ambitious humanists may push a cultural solution, e.g. a new philosophy, and underestimate its economic, ecological and political implications. The same could apply to any disciplinary expert, if they were to lack the time and incentives to estimate all derived implications on other fields, putting vested interests before the common good.

In this manner, disciplines may become part of the problem instead of the solution, leading to tainted policies, as in the example provided by Pontecorvo on the management of fisheries (Pontecorvo, 2003).

The unintended consequences of disciplines

The multi-faceted and complex nature of our social world leaves room for all disciplines and perspectives to understand it. Each of them prioritizes a specific aspect, leverages on a different epistemology, and relies – more or less explicitly – on a certain set of assumptions (Lele and Norgaard, 2005).

For this reason, I argue that each of them is to be considered partial and complementary to others, rather than exclusive or superior. Instead, within their logic and beauty, most disciplines and perspectives seem to give to their promoters good reasons to believe that they are superior to others, since "everything can be explained through them". In most cases, such overestimation may be exacerbated by mutual ignorance or by thinking to know more about "rival" perspectives than it is actually the case (Schroeder, 1997).

The science vs. God debate could be seen as a prime example of such misunderstanding: proponents of both perspectives seem to have sound reasons to believe that they are superior; both claim to know what the rival perspective is about and why; both appear convinced that their perspective cannot be reconciled with the other and exclude it a priori (Schroeder, 1997).

The attitude of mutual exclusion is subtly facilitated by language, in ways that we may not overtly be conscious of, most notably by preferring exclusive logical connectors, i.e. "or", "either…or", to inclusive ones, i.e. "and", "both…and". Such customary expressions may reflect a philosophically-rooted attitude to search reassurance in univocal truths, eventually leading to ideology if coupled with the human impulse to feel superior to others.

The very human (and in this respect legitimate and understandable) pursuit of self-reassurance and self-confidence may increase with the amount of time, energy and sacrifice that individuals dedicate to their disciplinary fields of expertise. Their personal and social identities rely on such commitment (Cerulo, 1997); deep satisfaction is derived from their beauty. Such emotional compensation lead the way towards wonderful and important disciplinary enterprises, but it may come at the expense of accepting other perspectives as equally important and deserving. The same may hold true not only for disciplines and epistemologies but also for professions, cultures and nationalities.

For instance: on which basis can we affirm that theory is superior to practice or rationality to emotionality? We would need to ground our reasoning on several assumptions or cultural axioms, prioritizing certain aspects of human behavior over others. This search for superiority may lead to misunderstandings and conflicts with other perspectives, which most likely are equally needed to understand reality, accept its complexity and eventually acknowledge our partial ignorance of it. On the contrary, humility and fraternity would be among the most important assets for the pursuit of interdisciplinary acknowledgement and mutual respect[2].

Psychological and ethical considerations aside, an additional downside of disciplinary overspecialization may be that of introversion and self-centeredness in the growth of knowledge if the search for problems and solutions is done exclusively within and for the discipline. We may end up unnecessarily overcomplicating reality, i.e. building a parallel reality from the reification of new theoretical entities, rather than making it more intelligible. The more time we spend going in depth within one particular perspective, the more we run the risk of isolating ourselves from the rest of the world, distracted from its primary and commonly accepted problems and unable to see the bigger picture.

Eventually, we may detach from a universal and normative common sense, which seems to coincide with those axioms on which all human beings in all cultures seem to agree, especially in regards to what is to be considered good and what is to be considered bad in human behavior (Emmons, 1972). As an example, we may hypothesize that some of the bankers and finance professionals held partly accountable for the recent global financial crisis were so self-absorbed by the beauty and power of their own money-maximizing algorithms that they overlooked the moral implications of what they were doing.

Such normative common sense may be then linked to what we call "common good", i.e. it may be precondition for it. If we detach from common sense, it may be more difficult to see the common good, i.e. what is universally considered good for humankind as

2 Granted the 2009 Nobel Peace Prize, US President Barack Obama may exemplify a new style of communication as here described.

a whole. Such universal common good may be grounded in those values which seem to be positively connoted across all cultures, such as life, justice, respect, natural environment, balance.

Last but not least, we need to acknowledge that another important set of reasons for people to advocate the superiority of a discipline over others is political in nature. Indeed, disciplinary research and speculation may overlap with partisan interests and interest groups on which they may depend for financial support and political influence. Economics is pushed forward by businessmen and entrepreneurs; the life sciences by the pharmaceutical industry; the humanities by artists and intellectuals; the engineerings by the military and the industry.

In some cases, such partisanship may be the ultimate reason to detach disciplines from the pursuit of the common good.

From insular to systemic interdisciplinarity

The issue of the insularity of scientific disciplines has been explored from different angles (Becher, 1990; Becher and Trowler, 2001; Overington, 1977). In order to publish in a given journal, the researcher most often needs to employ its terminology, cite previous work on the topic (Hamilton, 1990), as this is defined by the journal, and acknowledge (if not respect) what the journal sees as the most important theories and methods to understand the topic.

Journals are direct expressions of scientific communities, which by definition aim at defining who is to be considered part of the community vs. who is not. Among others, such insularity may serve two main roles: on the one hand, it grants the linear and incremental expansion of the knowledge base towards further specialization and detail; on the other hand, it confers a sense of social identity to the members of the community. Scientific communities are becoming more and more multi-disciplinary in that they invite experts from different disciplines to contribute to a given topic.

From a research standpoint (at least in the social sciences which I am familiar with) what is most problematic is that the topic itself dynamically changes as the result of complex forces of interdependence with other topics, and such interdependence is most often overlooked within the scientific community. For example, a journal in the field of transportation research could host articles related to the broad field of transportation written by a wide array of disciplinary experts: from geographers to economists, psychologists to engineers. Each would offer his or her perspective on the main topic; what I find to be missing in most cases is the acknowledgment and analysis of how the evolution of other high-order topics (i.e. society, economy, psychology) is affecting the topic itself (i.e. human mobility).

In other worlds, topics – and eventually disciplines – may end up being developed in a vacuum, as if they were not complexly intertwined with other topics or disciplines. When existing, the analyses of the impact of one topic on another takes place almost exclusively within individual papers oriented to single hypothesis testing (based on questions such as: "does more economic growth lead to more car traffic?"). Instead, very few journals focus on the evolving relationships between topics, e.g. – following our example – "Journal on the interdependence between the economy and transportation".

This problem is exacerbated by the fact that each topic is most certainly linked to more topics than one can imagine due to the higher and higher degrees of complexity and interconnectedness in today's global world. For example, transport and Chinese literature might look distant and unrelated, but only at a first sight: indeed, one could argue that Chinese literature contributed to today's Chinese culture in ways which in turn affect travel habits of Chinese people. In the same way, agriculture in Africa is affected by politics in the US, e.g. protectionist measures to shore off American agriculture from African competition, which in turn is affected by such disparate phenomena as Mexican immigration to the US, emerging agricultural technologies and the whims of global financial markets.

The same reasoning could also apply to the human body and the field of medicine. Different organs, the body and the mind, and the present and the past of a patient can all be considered discrete elements, yet they too interact in highly complex manners. Such interactions may make each disease manifestation unique (Gawande, 2003).

Most often, nevertheless, instead of integrating topics into their larger network, disciplines abstract topics from it, in order to allow further in-depth explorations. The manner in which they pursue interdisciplinarity, if they pursue it in the first place, is insular, as they approach a pillar topic *insulated* from complexity. Hence, beyond having different disciplines trying to explain the same issue from different angles – which could be defined as "insular interdisciplinarity" – I see the complementary need of explaining why a given issue is evolving as the result of interdependences with other issues – which could be defined as "systemic interdisciplinarity".

The pursuit of insular interdisciplinarity also applies to the business world. When designing a new product or service – for the sake of simplicity, a t-shirt – one needs to take into account its multiple roles: an ensemble of physical materials, an object of fashion, the result of a manufacturing process, an item to be sold at a price, a piece of clothing for people of different ages and cultures, and so on. The more the design process takes into account and mediates the contributions of different disciplines, e.g. material sciences, fashion, operational sciences and logistics, marketing, the more successful the final product is likely to be.

Nevertheless this is not enough, since selling t-shirts also has to do with the ability of matching them with other pieces of dress (i.e. trousers), the popularity of shirts as an alternative to t-shirts, shifts in the global economy that determine available expenditures for t-shirts, global fashions which determine the overall popularity of t-shirts, cheaper production modes such as outsourcing, and so on. As in the case of scientific research, systemic interdisciplinarity could be pursued by accounting for systemic complexity.

Conclusion

In this paper I have argued for the need of convergence of different disciplines for the common good. One of the most difficult steps for researchers and practitioners is the internalization of a new attitude toward knowledge creation: to recognize the equal importance and necessity of all disciplines, epistemologies and methods, and overcome psychologically and politically justified superiority or inferiority complexes.

Hopefully, a new awareness as well as the practice of common sense would limit a self-centeredness in the growth of knowledge which ultimately hampers the search for the truth in science, favors partisanship and polarizes discussion in the public opinion, as well as prevents comprehensive decision-making in politics and society. In this respect, further research is needed to explore how positivist scientific epistemologies can be reconciled with holistic and normative common sense.

I also argued for the need to complement what I defined as insular interdisciplinarity with systemic interdisciplinarity. If the former aims to approach a given issue from different disciplinary angles, the latter focuses on exploring how that given issue is the result of interdependence with other issues.

Ultimately, I see the pursuit of the common good as well as the practice of systemic interdisciplinarity as two fundamental steps to give humankind a better chance to address the biggest challenges of our time, from climate change to extreme poverty, from global security to population growth.

In parallel, the way forward should include a rethinking of the role of the individual, as well as introduce new incentives within society and its institutions. Not only to reward disciplinary expertise – which we still very much need – but also well-roundness and humility, empathy and wisdom.

References

Anderson, P. W. (1972) "More is different", *Science* 177, p. 393-396.

Becher, T. (1990) "The Counter-Culture of Specialisation", *European Journal of Education* 25, p. 333-246.

Becher, T. and Trowler, P. (2001) *Academic tribes and territories. Intellectual enquiry and the cultures of disciplines*, Buckingham, SRHE & Open University Press.

Cerulo, K. A. (1997) "Identity construction: new issues, new directions", *Annual Review of Sociology* 23, p. 385-409.

Emmons, D. (1972) "Normative Knowledge", *Journal of value inquiry* 6, p. 294-393.

Gawande, A. (2003) *Complications: a surgeon's notes on an imperfect science*, New York, Picador.

Hamilton, D. (1990) "Publishing by – and for – the Numbers", *Science* 250, p. 1331-1332.

Jervis, R. (1997) *System Effects: Complexity in Political and Social Life*, Princeton, Princeton University Press.

Lele, S. and Norgaard (2005) "Practicing interdisciplinarity", *Bioscience* 55, p. 967-975.

Overington, M. A. (1977) "The scientific community as an audience: toward a rhetorical analysis of science", *Philosophy and Rethorics* 10, p. 143-164.

Pontecorvo, G. (2003) "Insularity of scientific disciplines and uncertainty about supply: the two keys to the failure of fisheries management", *Marine Policy* 27, p. 69-73.

Schroeder, G. (1997) *The science of God,* New York, Broadway Books.

2.6 Frontiers Research: Seek, Share & Create

by Henry Markram and Kamila Markram[1]

In this paper, we argue that the ability to conduct advanced research and to access research performed by others are key to sustainable economic development. We go on to show that current subscription-based models of scientific publishing impose an impossible burden on developing countries that are cut off from key results, highly relevant to their development. We describe how Internet and Web 2.0 technology have enabled a new model of Open Access (OA) Publishing, providing free access to research results by scientists anywhere in the world. However, we also point out that OA publishing has failed to break the dominance of the traditional model. One of the reasons, we claim, is the way science is evaluated. After discussing and criticizing the weight currently assigned to Impact Factors, we go on to describe a novel system of peer review and "democratic tiered publishing" developed by Frontiers Research Foundation. Finally we discuss the relevance of this work to developing countries, taking the case of South Africa as an example, and showing how changes in publishing and review can be combined with other measures to promote science and innovation. We suggest that our work in South Africa could serve as a model for future work elsewhere in Africa and in other developing countries.

1 Henry and Kamila Markram are neurobiologists in the Life Sciences Faculty of the Ecole Polytechnique Féderal Lausanne (EPFL), where Henry Markram directs the EPFL's Brain Mind Institute. The two are also co-founders of the Frontiers Research Foundation, a swiss not-for-profit organisation supporting open-access knowledge dissemination.

Research Knowledge – a Foundation Stone for Development

Research knowledge is the foundation stone on which societies are built. Scientific knowledge comes from rigorous, reproducible experiments, based on rigorous standards defined and accepted by the world community of researchers. It is readily verifiable and easily transferred to future generations. It allows the creation of new technologies, new medical treatments, new agricultural techniques and new industries. It sets the tone of cultural and political discourse.

Academic papers shape future research, opening new prospects for society and setting policy agendas. Scientific publications make universities' reputations, giving them access to lucrative funding; scientists produce patents that drive the formation of innovative small and medium enterprises; royalties help the world's leading universities to grow and sustain themselves. In brief, countries that successfully generate world-class research develop economically and socially. Those that do not, slip into under-development.

It is a vicious circle. Academic research is expensive and the investment and pay-off is long-term, yet it is this research that drives education, innovation, invention, discovery and ultimately development. As countries fall behind in research, they find it ever harder to find funding and build their intellectual capital; top researchers move abroad; education standards drop. Without advanced academic research, they generate little intellectual property; when they need new products and services they have to license them from abroad.

Today's gold standard for the transmission of research knowledge is the peer-reviewed scientific paper. Every year scientists publish more than 2.5 million papers in some 24,000 scholarly journals. The first thesis of this article is that the subscription-based business models that currently dominate the market, fail to meet the needs of authors and readers, particularly in the developing world. We will outline the opportunities offered by Open Access publishing, arguing, however, that Open Access Publishing cannot be truly successful without changes in the way science is evaluated. We will therefore go on to introduce the innovative systems of review and open access publishing we have developed in Frontiers Research Foundation. Finally we will discuss how these innovations – combined with other initiatives – can contribute to development. As a case study, we will describe the current situation of research in South Africa, and work by Frontiers Research – South Africa to improve South African capabilities in research and innovation. In conclusion, we will suggest that a combination of Open Access Publishing, novel evaluation techniques, and active promotion of innovation can make a significant contribution to development, not only in South Africa but elsewhere on the continent and in the world.

The Price of Knowledge

Today, the main channel through which readers access research knowledge is through subscription-based journals from commercial scientific publishers. Over the last decade, this "pay to publish and read model" has come under intense critical scrutiny. According to the Organization for Economic Co-operation and Development (OECD) scientific research receives over US$ 2 trillion p.a. of tax-payers' money ever year. 80% of the papers it produces are captured and (re-) sold by subscription journals.

It has been claimed that the subscription journal is one of the most ingenious business models ever devised. Ownership of papers is vested in the publisher. Researchers spend taxpayer funds and student hours to generate research knowledge, just to hand over rights to access their work to a publisher. The results produced by a worldwide research budget of over US$2 trillion are essentially hijacked and held hostage by the subscription model. Researchers also provide journals with free peer-review services, often are asked to pay extra charges for publishing, and they must still buy access to their own paper. As authors compete with each other to give away their rights, journal prices rise and publishers continue to launch new journals for authors whose papers do not make it to top tier publications. The end result has been a continuous escalation of costs and a dramatic restriction on access. Between 1986 and 2003, US journal prices increased by 215%, compared to a rise in the Consumer Price Index of just 68.9% (Albert, 2006). This was what has been called the "serials crisis". As a consequence, US research libraries spent 2.7 times more for serials in 1998-1999 than in 1985-1986, while still purchasing 6% fewer titles. According to Steven Harnad, a leading proponent of "Open Access",

> *"Even Harvard, the university with the largest journals budget of all, cannot subscribe to all journals, and most other universities have far smaller journal budgets".*

The problems facing researchers in developing countries are even more severe. A study by the WHO, in 2003, showed that 56% of medical institutions in 75 countries with GDP per capita per year of less than US$1,000 did not have a subscription to a single medical or scientific journal (Aronson, 2004). In practice, the serials crisis restricts access to research knowledge to the 10% of richest countries in the world. Managing the submission and review process, processing articles for publication, printing and distributing a high quality journal is expensive and publishers obviously wish to recoup these costs. But the business model they have invented limits access to knowledge for everyone. It is legitimate to ask whether this kind of restriction can be ethically justified in the information and knowledge societies of the 21st century.

Open Access Publishing

The Computer and Internet Revolution of the 20th century set the scene for the Information Revolution of the 21st. Tim Berners-Lee's[2] invention of the World Wide Web and the subsequent emergence of Web 2.0 marked the beginning of a new era of pervasive information, available at all times, free of charge and from anywhere in the world. This was a very different paradigm from the restricted access to knowledge offered by subscription-based scientific publishing.

The new technologies spurred libraries and researchers from physics, mathematics, medicine and neuroscience to create the Open Access Publishing movement. The result

2 Inventor of the World Wide Web in 1990 and current Director of the World Wide Web Consortium.

was a new model of publishing offering scientists a service that was "*digital, online, free of charge, and free of most copyright and licensing restrictions*"[3].

The years from 1991 to 1998 saw the creation of the first freely accessible online repositories of research articles, among which, ArXiv (www.arxiv.org). The launch of ArXiv's e-print archive server in 1991 was rapidly followed by the emergence of Open Access advocacy groups, such as SPARC (www.sparc.org), and the creation of large public resources such as the Public Library of Science (PLoS, 2001; www.plos.org) and PubMed Central (2002; www.pubmedcentral.nih.gov). Today, more than 4,000 OA journals[4] are beginning to exert a significant impact on citations and downloads.

According to Alma Swan, a prominent open access advocate,

> "*Open access means there is greater visibility and accessibility, and thus impact from scholarly endeavor; open access means there is more rapid and more efficient progress of scholarly research; open access means there can be better assessment, better monitoring and better management of science, and finally, open access means that novel information can be created using new computational technologies.*"[5].

John Willinsky, Professor at the Stanford University School of Education and an Open Access activist argues that the scholarly community has a moral obligation to disseminate research to the widest possible audience. In his words,

> "*A commitment to the value and quality of research carries with it responsibility to extend the circulation of such work as far as possible and ideally to all who are interested in it and all who might profit by it.*" (Willinsky, 2006)

It is this motivation that has driven the global Open Access movement. As Heather Morrison, project coordinator of the British Columbia Electronic Library has put it:

> "*An old tradition and a new technology have converged to make possible an unprecedented public good (Budapest Open Access Initiative). The old tradition is that of scholars giving away their peer-reviewed research articles. The new technology is computers and the Internet.*"[6]

OA publishing is also receiving increasing support from the international community. In 2002, UNESCO launched *Information for All*[7], an intergovernmental program to facilitate universal access to information and knowledge. In 2003, the Berlin Declaration

---

3 See online article by Peter Suber; Open Access Overview: Focusing on open access to peer-reviewed research articles and their preprint: http://www.earlham.edu/~peters/fos/overview.htm

4 4,141 according to the latest updates on the Directory of Open Access Journals.

5 http://www.keyperspectives.co.uk/openaccessarchive/Journalpublications/Belgian%20library%20journal%20article%20%20final%20revised%20version.pdf

6 http://76.12.54.203/work/openaccess/assets/Open%20Access%20Introduction.pdf

7 http://portal.unesco.org/ci/en/ev.php-URL_ID=1627&URL_DO=DO_TOPIC&URL_SECTION=201.html

on Open Access to Knowledge on the Sciences and Humanities[8], and the Bethesda Statement on OA Publishing[9] reiterated international support for the goals of the OA movement. On January 30, 2004, the OECD Ministers for Technology and Scientific Research endorsed the Declaration on Access to Research Data from Public Funding[10], calling on governments to guarantee taxpayers' right to access scientific knowledge. The Declaration of Principles issued by the 2005 World Summit on the Information Society committed the signatories:

> "... *To promote universal access with equal opportunities for all to scientific knowledge and the creation and dissemination of scientific and technical information, including open access initiatives for scientific publishing*"[11].

And yet, despite these efforts, Open Access continues to face at least two critical challenges. The first is the strength and reputation of subscription journals, which continue to hold an 80% share of the scientific publishing market. Recent reports by the German Research Foundation and the UK JICS Scholarly communication group suggest that, at least in some disciplines, scientists are still largely unaware of OA publishing. Despite studies showing that papers in OA journals have higher "visibility" than papers in subscription publications, many researchers still believe that the best way of promoting their careers is to publish in traditional, high impact journals. Realism suggests that at least for young researchers, publishing their work for the first time, this conviction may be correct.

The second challenge is that OA publishing on its own does not solve the access problem for developing countries. Indeed, the first effect of the new model may well have been to widen the gap between North and South rather than to narrow it. According to Subbiah Arunachalam, a Chennai, India, based information consultant, and a strong advocate for Open Access:

> *"It is in the nature of any new technology to exacerbate the existing divide between the rich and the poor. The newer and more potent the technology, the greater its ability to increase the inequalities."* (Arunachalam, 2003)

Arunachalam is concerned that poorer or disadvantaged regions have only a limited ability to benefit from innovation, even when the innovation itself is positive. Even with Open Access Publishing, developing countries remain decades behind the curve in their ability to produce and profit from world-leading research.

8 http://oa.mpg.de/openaccess-berlin/berlindeclaration.html
9 www.earlham.edu/~peters/fos/bethesda.htm
10 http://archiv.twoday.net/stories/133859/
11 UN meeting urged to back open access science, David Dickson, 7 December 2003: www.scidev.net

OA Publishing and the Developing Countries

In general OA has progressed only slowly in the developing countries. Nonetheless, a number of government and NGO initiatives are already underway. Some of the most important are in India, Brazil and South Africa.

In India, workshops and seminars organized by the Indian National Science Academy (INSA) in 2003, and by the Indian Academy of Science in 2006, have improved awareness of OA and open archiving. One of the outcomes of the 2006 meeting was the Bangalore Model National Policy Statement. The Statement,

> *"Provides a clear way forward to achieving free access to publicly-funded research publications that is essential for scientific progress in all countries. It can be adopted and used by national governments, their funding organizations, research institutes and universities to accelerate the free exchange of research findings and reap optimum benefit from academic investment."*[12]

A number of services are facilitating the use of OA to distribute results from Indian research. One of these is Medknow Publications, a company based in Mumbai that publishes 59 medical journals on behalf of Indian institutions and associations. A recent decision by Council of Scientific and Industrial Research (CSIR) made OA research publishing mandatory[13] for government-funded research. This is obviously a major step forward.

Latin America and Brazil, in particular are also embracing OA. Today, most research output from Latin America is distributed via OA journal services, such as Bioline International a Brazil-Canada initiative that provides assistance to publishers in developing countries, managing a collection of over 70 OA journals published in 17 countries. SciELO (Scientific Electronic Library Online), managed by a large group of Latin American publishers, publishes over 500 journals, in Brazil, other Latin American countries, Spain and Portugal[14]. At the 9th World Congress on Health and Information Libraries held in Bahia, in 2005, Brazil proclaimed the Salvador Declaration on Open Access, calling on *"governments to make Open Access a high priority in science policies,"* and appealing for *"all stakeholders in the international community to work together to ensure that scientific information is openly accessible and freely available to all, forever."*[15]

On the African continent, it is South Africa and Nigeria that lead the field. In South Africa the position of OA has been helped by successful initiatives such as Sabinet's Open Access Journal Collection and African Journal Online, a South-African based non-profit organization that provides online access to African-published research, and hosts over

---

12 Arunachalam on the Bangalore Policy Statement: Workshop on Electronic Publishing and Open Access Indian Institute of Science, Bangalore, November 2-3, 2006.

13 Sreelata, M. (March 13, 2009). Key Indian research organization goes open access. http://scidev.net/en/science-communication/open-access/news/key-indian-research-organisation-goes-open-access.html

14 The problems faces by research communities in developing countries: http://www.openoasis.org/index.php?option=com_content&view=article&catid=78:articles&id=28:developing-countries&Itemid=253

15 Salvador Declaration on Open Access: the developing world perspective, Salvador, Bahia, Brazil, September 23, 2005.

340 African-published, peer-reviewed journals from 26 countries[16]. The Academy of Science of South Africa (ASSA) has recently started a new project based on the example of SciELO in Brazil. The South African Scientific Journal, published by the Academy, is the first high-profile OA journal in Africa.

On top of these regional initiatives, International organizations are engaged in a major effort to improve developing country access to science, technology and medical research. In 2002, for example, the World Health Organization launched The HINARI (Health InterNetwork Access to Research Initiative) program, providing researchers in developing countries with free or reduced rate access to over 2,000 medical, biomedical and social sciences journals (Winterbottom, 2006). A second, similar program, AGORA, focuses on agricultural research.

Programs like these tend to attenuate the access crisis for developing countries, but they do not solve it. According to Laura Wimberley,

> *The "HINARI program has been a disappointment so far…researchers in Peru for example were unable to access most of HINARI journals with the highest impact, and the interface is difficult to search with problems when trying to access it."* (Wimberley, 2008)

All these programs are restricted to developing countries with per capita GDPs of less than US$1,000, excluding countries such as India, where per capita GDP has reached US$3,000. Ultimately, they do little more than provide artificial life support for a dying model of science publishing. Later in this paper we will outline practical strategies to improve access to research knowledge in developing countries and describe our attempts to implement this kind of strategy in the setting of South Africa. But before these strategies can work, other changes are required. One of the most urgent needs is to change the way governments, universities and journals assess the quality of scientific research.

The Impact of the Impact Factor

> *"… As long as it is more important where an article is published than what it is about and how it contributes to scientific knowledge, we will continue to have problems…"*
>
> Peter A. Lawrence, *Nature* 2003, 422, p. 259-261

In the last analysis, the continued dominance of subscription-based publishing depends on researchers' need to "publish or perish" or more specifically on their need to publish in journals "of high repute". The main way of measuring repute is the *Impact Factor.*

September is a worrying time for publishers and scientists. This is when the Institute of Scientific Information (ISI), publishes its annual list of Impact Factors. The list ranks

16 http://ajol.info/index.php/ajol/pages/view/AboutAJOL

more than 6,000 journals based on the frequency with which they are cited in other journals. Many scientists' careers depend on the ranking.

Citation indexing began in the 1950s, driven by dissatisfaction with traditional ways of managing information and the hope that automation could provide the answer[17]. The new technique made it possible to retrieve relevant literature across arbitrary classifications in a way that subject oriented indexing could not. Many of its proponents hoped it would also eliminate subjective human judgment in the selection of descriptors and indexing terms.

In the early 1960's, Eugene Garfield and Irving H. Sher created the journal *Impact Factor*. The US government had chosen not to subsidize a national citation database, so Garfield set up a private company to publish the index. The name of the company was Thomson Scientific. Later Thomson was to create ISI, the Institute for Scientific Information. It was ISI that was charged with indexing journals and calculating Impact Factors.

A journal's Impact Factor (IF) is computed as the number of citations received by the papers it has published over the previous two years, divided by the total number of "citable" articles it published in the same period (PLoS Medicine (Eds), 2006).

In recent years, the procedures used to calculate the index have been subjected to heavy criticism. Doubts raised by critics include the following.

1. A journal's Impact Factor is based on the overall number of citations of all the articles it publishes; the IF has nothing to say about any specific article, let alone about a particular author.
2. It is ISI that decides what articles should be considered as "citable". This decision can substantially change a journal's IF. Similarly it is ISI that decides which journals to include in the IF list. According to ISI, only 10% of the 2,000 new journals reviewed annually, are selected (Belew, 2005).
3. The most frequently cited articles are review papers. As a result, journals that only publish review articles have an unduly high IF.
4. IFs say nothing about an article's impact outside the scientific community, for instance when it is used as the basis for a new product or technical method.
5. Famous scientific errors and "worst papers" often attract many citations.
6. IFs are subject to manipulation via self-citation. In 2007, a journal protested against the absurdity of IFs with an editorial citing all its articles from 2005 to 2006. Its IF increased.
7. A journal's IF is determined by its editorial policies rather than by the value of the papers it publishes.
8. In most fields, 20% of the articles published account for 80% of all citations.
9. IFs measure popularity rather than prestige.
10. ISI does not index citations in books. This means that IFs may be relevant for science and economics, where most citations appear in journals, but not for the humanities where they often appear in books.

Jennings (Jennings, 1999) has argued that authors, choosing a journal for a paper, rely too strongly on Impact Factors rather on than more appropriate criteria, such as edi-

[17] History of Citation Indexing, http://thomsonreuters.com/business_units/scientific/free/essays/history

torial practice or target readership. Worse, Impact Factors are currently used, not just to measure the repute of a journal, but to assess the impact of individual articles and authors. This may well have inappropriate effects on scientists' careers, institutions, and grants. As even Eugene Garfield, ISI's founder, admits:

> *"The Impact Factor was originally intended as an objective measure for the reputability of a journal, it is now being increasingly applied to measure the productivity of scientists."* (Smith, 2008).

Bollen *et al.* (2008) have shown that scientific impact is a multi-dimensional construct that cannot be adequately measured by any single indicator. A principal component analysis on 39 measures of scholarly impact, half based on social networking data and half on Web access logs, showed that the most reliable measures of scientific prestige were the measures based on usage. Interestingly rankings based on different measures all correlated well, with the single exception of IF. In brief, today's most widely used technique for measuring the impact of journals is itself unscientific and highly subjective. Against this background, it is not surprising that several governments have recently issued guidelines, requiring universities to use experts to evaluate publications by job candidates rather than relying on IFs. According to Wolf Singer, editor of *Frontiers in Cellular Neuroscience*, this represents a major cultural change.

Yet the fact remains that scientists still seek publication in subscription journals with high IFs and that many of the important results from scientific research are available only to those who can pay to read them. This usually means researchers in rich countries.

A New Approach to Review and Evaluation

Recent years have seen many attempts to break Thomson's monopoly on measures of scientific impact. Some are based on citations (Google Scholar, CrossRef, Hirsch(H)-Index, PageRank), some on usage (e.g., Y factor), some on numbers of downloads, page views etc. None are entirely satisfactory.

The Frontiers journal system, which is a grass roots initiative "by researchers for researchers" is preparing to launch two new indices, which we have called the Index of Academic Excellence and the Index of Social Relevance. Unlike IF, the new Frontiers indices are deliberately designed to measure the impact of ideas, articles and authors; unlike IF, they allow readers and users of data to share index values and unlike IF, they measure the impact of research not only on a specific discipline but on the broader scientific community and on society. The goal is to move beyond a system that measures "where you publish" to one that assesses "what you publish".

Yet to achieve this goal, indices on their own are not enough. Frontiers has thus introduced two additional innovations that we believe to be of key importance. First we have introduced a radically new review process; second we have introduced a system of tiered publishing that for the first time gives a direct role to readers in assessing research impact. It is these innovations that we will now proceed to discuss.

The Frontiers Review Process – Evaluating the Quality of Research Knowledge

Frontiers believes that the best strategy for ensuring the quality of papers is truly collaborative interaction between authors and reviewers. To this end, Frontiers has re-designed the traditional peer-review system to ensure that articles are assessed in a fair, constructive and unbiased manner and that only the best reach publication.

The new Frontiers review process is based on a strict separation between review and evaluation. Review editors can reject a paper only when it contains objective errors. Their job is to ensure that experiments are valid, results flawless, and quality is as high as possible. Evaluation of significance and importance is left to the Frontiers tiered publishing system, which we will discuss in the next section of this paper.

The review system is based on seven key innovations.

- Prior to publication of a paper, Frontiers publishes an "intermediate version", providing proof of priority for discoveries still in the review process.
- Review editors limit their evaluation to objective judgements on the design and implementation of experiments; subjective judgements on impact are not allowed.
- Frontiers offers authors and review editors a real-time interactive forum where they can iteratively question the papers vailidity and discuss and improve a paper's content and presentation.
- Frontiers provides an arbitration process to settle disputes between authors and review editors.
- Acceptance or rejection of a paper requires the unanimous agreement of the review editors – no more voting system to judge true and false.
- Published papers disclose and acknowledge the names of review editors and associate editors making reviewers publicly responsible for the review process while also recognizing their contribution.
- Evaluation of subjective issues of impact is performed democratically, only after publication by assessing the readership of the paper.

Frontiers reviews consist of two stages: independent review and interactive review.

Independent review

When an article is submitted, the associate editor scans the article and decides if it is of sufficient quality to merit review. In this case, the associate editor assigns a minimum of two review editors to the paper. All review editors are scientists with an outstanding record in their respective fields of specialization. When an associate editor believes an article does not merit review, he or she recommends rejection. The final decision to reject belongs to the specialty chief editor who has the right to override the recommendation of the associate editor.

Once a paper has been sent for review, invited review editors are obliged to accept or decline the invitation within two working days. If no reviewers want to review the article, this is grounds for rejection. When an invited review editor accepts a review invitation, he or she has ten days to prepare an "independent review report". Each review editor works independently of the others. All reviewer recommendations are included in the independent review report, which is based on a standardized report template. The use of the standardized template ensures equal, fair, rapid, comprehensive and comparable assessment of research. Frontiers procedures do not allow accompanying letters to the associate editor, or hidden recommendations to accept or reject a paper.

Interactive review

Once the independent review reports have been received, the associate editor activates the "real-time interactive review forum" – similar in many ways to an Internet discussion group. The forum is activated, even when independent review reports are unfavorable to the authors. This gives authors an opportunity to rebut criticisms from review editors.

All participants in the forum (authors, review editors, the associate editor, the chief editor) can view each other's reports, replies and comments, and interact with each other in real-time. Authors can engage in multiple iterations with review editors and resubmit revised versions of their manuscripts. This process allows a convergence of opinions and improves the general quality of papers. Where authors and editors cannot reach agreement, each party expresses its opinion, after which authors can resubmit a new version of the paper. Unlike other publishers, Frontiers places no limit on the maximum number of resubmissions.

The interactive review forum is designed to achieve consensus on all issues involving objective errors. In the event of a dispute, authors or review editors can request arbitration. The initial arbitrator is the associate editor of the journal, who will discuss the disputed issue with the review editors and attempt to resolve the problem. If this does not work, the specialty chief editor is informed. He or she can then bring in additional review editors or associate editors for consultation on specialized topics. If the arbitration rules that a paper does indeed contain objective errors, it can be rejected.

The review ends when both review editors consider that their concerns have been satisfactorily addressed. Given that their names will appear on the published paper, it is important that they should be genuinely satisfied. In this case, they finalize their reports, signalling to the associate editor that the paper is ready for acceptance – unanimity to accept, unanimity to reject. The associate editor has sole responsibility for the final decision to accept a paper. The chief editor is not required to approve the decision.

If both review editors agree that an article has insurmountable problems or objective errors, which the authors are unable or unwilling to address, the associate editor may recommend rejection or invite a new set of review editors. In cases where the recommendation is rejection, the specialty chief editor enters the interactive review forum, examines the reviews and interactions and takes the final decision. When articles are rejected, the review editors remain anonymous. This ensures that editors can be as critical as they wish during the review process and protects their right to withdraw from the review process.

If a review editor is unsatisfied with the review process, he or she may withdraw and remain anonymous. In this event, the associate editor invites another review editor. The frontiers system guarantees equal right for authors, reviewers, editors and readers.

The Frontiers Tiered Publishing System – Evaluating the Significance of Research Knowledge

The Frontiers review process is designed to ensure that papers published by Frontiers journals are of the highest possible quality and that they contain no objective errors. Yet, high quality, on its own, is not enough to meet societal needs or those of the scientific community. In 2008, journals listed by ISI published over 1.5 million scholarly papers and unlisted journals published many more. Despite the vital work of science journalists and popular science authors, many important results failed to be brought to the public or broader scientific community's attention – a risk which is particularly high when papers are published in less well-known journals or when they discuss issues of prime concern to developing countries. In other words, there is a desperate need for an unbiased global system, capable of identifying major breakthroughs with a high potential impact on science and society. This is what we are trying to achieve in the Frontiers Democratic Tiered Publishing System.

The goal of the system created is to automatically and democratically evaluate the scientific excellence and social relevance of research ideas and articles, whose general validity has already been veted by the Frontiers review process. An academic excellent index depends on the impact of a published paper on the expert community from which the paper came, while the social relevance index depends on the impact of the paper on the wider scientific community and eventually even on the public.

The Frontiers system was launched in 2008, using neuroscience to prototype the system. Within 2 years Frontiers became the 2nd to 3rd largest publisher of neurocience knowledge in the world. The full Frontiers system, still under development, will eventually be organized into four "tiers": at the bottom, articles are published in approximately 2,000-3,000 specialty sections as 1st tier articles, highly specialized and expert-oriented research result presentations. The best articles – as evaluated democratically – "climb the tier" to become prestigious 2nd tier publications in one of the 120 field journals. The best articles are again selected to become highly prestigious 3rd tier features in one of the 5 domain journals, addressing the general public. The top 4th tier consists of a single unifying magazine, designed to present results from lower tiers to the general public; a single research knowledge platform for all of academia.

When Frontiers accepts a paper it is published online in the appropriate specialty section. The Frontiers software platform tracks the activity of readers who access the journal, using the data to compute indices of academic excellence and social relevance. Every three to four months, the top 10% contributions are selected. The authors of these "top ten" articles are invited to write a review-style paper (a "Focused Review"), positioning their original discovery in a wider context and pitched to the broader scientific community – the field. Focused Reviews are centered on the wider implications of their original discovery, and are just a few printed pages. All Focused Reviews are edited for layout and readability, guaranteeing they are accessible to a wider audience.

The next 10% best discoveries become highlighted in Frontiers Commentaries, written by the Associate Editor(s) who handled the article during the review process – taking on a role similar to that of scientific journalists. Like Focused Reviews, these articles are published in 2nd-tier publications and go through the normal Frontiers review process prior to publication.

As Frontiers expands its activities to cover an ever broader range of scholarly disciplines, the social relevance index will be become more important than the academic excellent index in driving the selection of articles up to the third tier covering science, medicine, technology, society and culture and finally to the Frontiers Magazine. It goes without saying that the papers that appear in the Frontiers Magazine will be those that the academic community has judged to be excellent and which potentially millions of readers have considered to be of high social relevance.

Frontiers' software platform tracks reader access to abstracts, full-texts and PDF files, measuring how often an article is read, how long readers spend on the article and even the location of the reader. The scores of individual articles are updated dynamically depending on clicks and downloads, e.g., number of abstract views, number of full text views, and number of downloads.

After 2 years of experience with the system and much feedback from users, Frontiers has launched a new improved analytics platform to reflect the impact of a new article to the readers in real-time. In the new system, scores take account of *who* reads *what* and to what depth. Data from individual readers are normalized to give more weight to readers with higher expertise when generating the academic excellent index, and more weight to those experts from far afield when generating the social relevance index. The new Frontiers Evaluation System is the first in the world to allow all readers to participate in the evaluation of research articles and the first to provide a normalized democratic assessment of the relevance of scientific research to society.

Frontiers South Africa

The African Challenge

Nowhere are Open Access publishing and new methods of evaluation more important than in Africa. The United Nations, the European Union, the World Bank, foundations and private groups have launched a myriad of initiatives to reduce child mortality, improve maternal health, combat HIV, malaria and other diseases, support social, political and environmental sustainability, and develop a global partnership with Africa (Cokerill and Knols, 2008). However, it is hard to believe that these initiatives can be successful when many African institutions and leaders remain unaware of the latest advances in scientifically-validated social, political, scientific, medical and technological research. So long as Africa has to import knowledge and technology, it will need to be fed and supported from the outside.

Research is the essential precursor to technological innovation, and to social, political and economic stability. The emergence of Africa will ultimately be measured by its ability to sustain itself and to create and export its own knowledge and technology. Yet without access to the latest research knowledge, Africa cannot expect to generate intellectual property of strategic value on a highly competitive world market. This requires Open Access Publishing and methods of evaluating research than can identify results relevant to African needs. And it needs leadership from within the region. Perhaps the strongest candidate for this role is South Africa.

The South African Research Knowledge Crisis

A first visit to South Africa surprises even seasoned travellers. They discover six lane highways, hotels and restaurants well above the standards of the western world, a strong appreciation for nature and wildlife, an impressive mix of modern and colonial architecture and much else besides. South African universities, modelled after the British system, have produced a strong research, medical and technical community. South Africa produced the world's first heart transplant, some of the most advanced mining technology, and advanced marine biology strategies to preserve the oceans wildlife. Yet despite these achievements, current prospects for the future of South African science are bleak.

There are many ways of measuring a country's status as an advanced knowledge society. Key indicators measure the proportion of GDP invested in research and development, the number of universities and researchers, the research articles they publish, the patents they file, and the journals they can access. We can use these indicators to outline the current situation of South African research.

Research & Development Expenditure

South Africa spends just 0.9% of GDP on research and development.[18] This is not only lower than the equivalent figures for the USA, Europe or Australia; it is also lower than the figures for China and India.

Number of Science and Technology Researchers

South Africa has just over 17,000 researchers – just 0.03% of the total population. This is roughly the same proportion of the population as India or Brazil – higher than the figures for other African countries, but much lower than those for the USA, the European Union or Austrãlia. According to a recent UNESCO[19] report, South Africa has 24 public and private universities and a total of about 750,000 students – equivalent to roughly 1.5% of the population. The situation in India and China is similar; both have about 1% of the population in university education. However, the South African student population is much smaller than in European transition countries such as Poland and Russia, where students make up 5.4% and 5% of the population, respectively. In the older industrial countries, the proportion of the population attending university ranges from 2.3% in Germany, and 3.6% in the Netherlands. In highly developed Scandinavian countries such as Sweden and Finland, the figures are even higher (4.4% and 6.0%, respectively).

Number of Research Papers

Table 1 shows the number of publications published by South African researchers during the two 10 year periods ending in November 1994 and November 2000. Overall, South Africa contributes less than 0.2% of the 2.5 million research articles published every year. Worse, research output is growing around 10% slower than output from the rest of the world (Pouris, 2006).

---

18 Gross domestic expenditure on R&D (GERD) as a percentage of GDP, 2005 or latest available year. Source: UNESCO Institute for Statistics, September 2007.

19 UNESCO Institute for Statistics, September 2007.

Table 1 South African share of research publications by major disciplinary area. The share is falling in all areas.

Discipline	1990-1994	1996-2000	Growth/Decline (-)
Life Science	0.45%	0.39%	−13.3%
Traditional sciences	0.35%	0.29%	−17.19%
Technology	0.40%	0.38%	-5%
Land and primary	1.32%	1.2%	-9%
Social Sciences and humanities	1.42%	0.5%	−19%
Multidisciplinary	1.25%	1.16%	-7.2%
Average	0.56%	0.50%	−10.7%

South African Access to Journals

South African universities have access to less than 2% of the 24,000 academic journals and 212,000 periodicals published in the world (Figure 1).

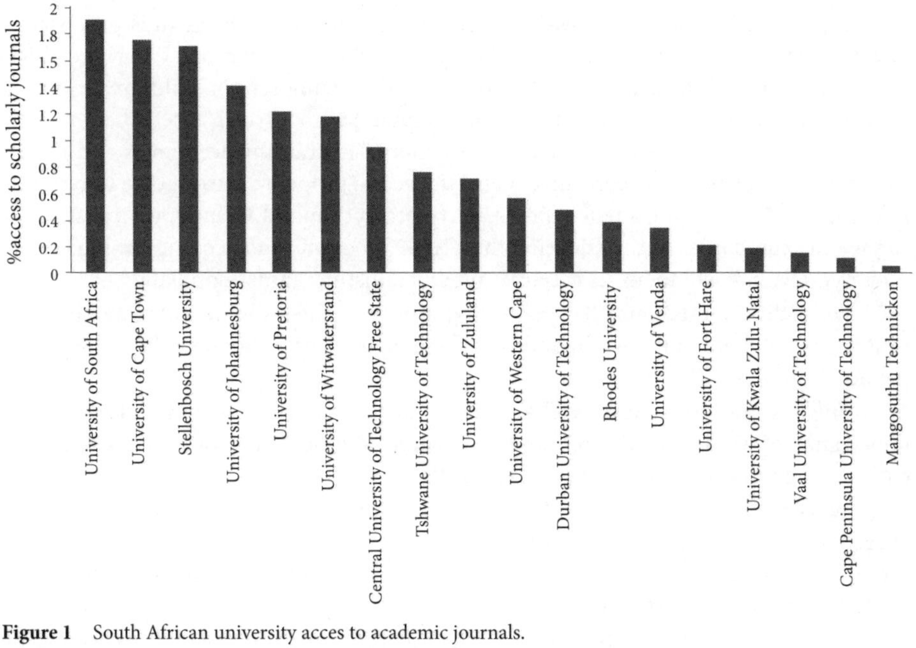

Figure 1 South African university acces to academic journals.

Journal Publishing

Another indicator is the number of journals published in South Africa and the number of Open Access journals launched by South African institutions. According to the latest statistics from the Directory of Open Access Journals, only 87 of the 24,000 scientific journals published in 2009 came from South Africa. Of 4,352 Open Access Journals, just 14 were South African. South Africa hosts only 0.06% of the academic journals published in the world.

Intellectual Property and Patents

In 2006, South Africa contributed 874 of 1.8 million patents filed worldwide. In brief, South Africa generates just 0.05% of the world's intellectual property[20].

Frontiers Research Foundation – South Africa

South Africa has the potential to become a 21st century knowledge society – a world leader in science, medicine, technology, engineering, business studies, economics, the social sciences, humanities, and the arts. But to achieve this goal South Africa needs to be as competitive in innovation as it already is in sport. This is what we are trying to achieve in Frontiers Research Foundation South Africa (FRF-SA). FRF-SA aims to ignite the spirit of innovation in the minds of South Africans, young and old; encourage and support innovation in South African universities; and build up the innovative capabilities of South African companies. In so doing it can provide a model for the rest of Africa.

South Africa has more natural resources than many other countries, but is years behind the curve in the race for knowledge and innovation. FRF-SA believes that the best way of developing South African research capabilities is to let loose South Africans' natural spirit of invention, innovation and competitiveness. FRF-SA thus proposes an ambitious systems program aimed at building up public understanding of advanced academic research and its importance for society, encouraging school children and village people to innovate and to make inventions, helping South African researchers to compete with cutting edge research in advanced countries, training academics and professionals to protect and commercialize their intellectual property, introducing universities to the latest strategies in intellectual property protection and technology transfer, and supporting companies as they develop their R&D programs and accumulate intellectual property. FRF-SA will focus its activities on seven distinct target populations.

Public: the Foundation will support programs to spread awareness of research results generated elsewhere in the world and to make them easier to understand for the general public.

Children: the Foundation will support activities exposing school children to the importance of academic research and innovation for the future of South Africa, and encouraging innovation, discovery and invention.

Students: the Foundation will support programs to help South African students enaged in the intensely competitive international arena of generating research knowledge through exposure to international standards, the latest research strategies, and methods of scientific writing and presentation.

Researchers: the Foundation will support programs to improve South Africa's access to advanced research knowledge, fostering international collaborations, helping researchers to gain access to international funding, and providing guidance to academic researchers seeking to protect their discoveries and intellectual property.

Universities: the Foundation will help South African universities realize the enormous value of investing in research, giving them access to the latest strategies in technol-

[20] World Patent Report (2008) – A Statistical Review (http://www.wipo.int/export/sites/www/ipstats/en/statistics/patents/pdf/wipo_pub_931.pdf).

ogy transfer and the marketing of research results, and assisting them with strategies to compete for international research grants.

Professionals: the Foundation will bring professionals into contact with the latest research in their area of specialization, showing them how to register and exploit intellectual property.

Companies: the Foundation will make companies more aware of modern strategies for designing and managing R&D programs, accessing international R&D funding, securing intellectual property and turning research results into fruits for society.

Conclusion

The central theme of this paper has been the fundamental contribution of scientific research to development. Without the ability to conduct their own high level research and without access to research results from elsewhere, countries have little prospect of achieving sustainable economic or social development. We have also seen that the way in which research is published and assessed actually work against accessibility. The continued dominance of subscription-based publishing models imposes an impossible economic burden on developing countries; over-reliance on Impact Factors perpetuates the model, while failing to identify ideas and articles of high relevance to their needs. For Frontiers, accessibility of research results and proper assessment of research are moral imperatives. In the same way, we believe it is a moral imperative to promote scientific research in South Africa and elsewhere in the developing world. In this paper we have outlined how Frontiers is trying to respond. Obviously we do not believe we have all the answers but we do believe that some of our activities can begin to serve as a model for future work elsewhere in Africa and the developing world. Our hope is that others can learn from and improve on what Frontiers is trying to achieve.

Acknowledgements

Materials and sources were obtained by interviews and research by Frontiers Research Foundation (Switzerland). In particular the work of Shamsa Abdulrasak, Nathalie Tzaud, Anna Muller, and Nicole Detzer from the Foundation. Frontiers thanks Richard Walker for his valuable editing of this article. Frontiers Research Foundation is supported by the AVINA Foundation in Switzerland and a number of visionary philanthropists.

References

Albert, K. (2006) "Open access: implications for scholarly publishing and medical libraries", *J Med Libr Assoc.* 2006 July, 94(3), p. 253-262.

Aronson, B. (2004) "WHO survey: Access to medical information for low-income countries", *NEJM*, 350, p. 966-968.

Arunachalam, S. (2003) "Information for Research in Developing Countries: Information Technology – Friend or Foe?", *Bulletin of the American Society for Information Science and Technology*, Volume 29, Issue 5, p. 16-21.

Belew, R. K. (2005) *Scientific impact quantity and quality: Analysis of two sources of bibliographic data*, arXiv:cs/0504036v1[cs.IR].

Bollen, J., Van de Sompel, H., Hagberg, A., and Chute, R. (2009) *A principal component analysis of 39 scientific impact measures*, PLoS One 4. doi:10.1371/journal.pone.0006022.

Cockerill, M.J. and Knols, B. G. (2008) "Open Access to Research for developing world", *Issues in Science and Technology*, The University of Texas at Dallas.

Jennings, C. (1999) "Citation data: the wrong impact?", *Neuroendocrinol.* Lett. 20, p. 7-10.

PLoS Medicine (Eds) (2006) *The impact factor game*. PLoS Med3,doi:10.1371/journal.pmed.0030291.

Pouris, A. (2006) "Local Science needs a booster shot", *Essays Innovate* N° 1, p. 90.

Smith, R. (2008) "Beware the tyranny of impact factors", *Journal of Bone and Joint Surgery – British Volume*, Vol 90-B, Issue 2, p. 125-126.

Willinsky, J. (2006) *The Access Principle: The Case for Open Access to Research and Scholarship*, by, The MIT Press, p. 307.

Wimberley, L. (2008) *Open Access Journals in the Developing World*. Open and Libraries Class Journal, Vol 1, N° 1 (2008) http://infosherpas.com/ojs/index.php/openandlibraries/article/view/6/11

Winterbottom, A. (2006) *Open Access: scientific publishing and the developing world*. Oxford, First Author. http://www.firstauthor.org/Downloads/openaccess.pdf

Chapter 3

Common Language

3.1 Towards a Science of Collective Intelligence

by Pierre Lévy[1]

How can the ubiquitous resources of communication and computing offered by the digital medium be fully exploited in modeling and augmenting human collective intelligence? This is the question where my research originates. The expression "collective intelligence" can have many different meanings, but all these meanings are more or less linked to the convergence of two concepts: cognition («intelligence») and society or community ("collective"). Cognition refers here, very classically, to the activity of perceiving, remembering, problem solving, learning, etc. «Collective intelligence» thus refers to the cognitive abilities of a society, community or collection of individuals.

Starting from this very general approach, collective intelligence can be considered from four different but interdependent points of view: firstly, the collective intelligence of animal societies; secondly, that of human societies in general; thirdly, the collective intelligence the growth of which constitutes an explicit goal in knowledge societies that cultivate a new "information economy"; fourthly, the collective intelligence of those communities that self-organize through the digital medium. The present paper will focus above all on integrating the last three points of view using a scientific model of collective intelligence: the IEML (Information Economy Meta Language) semantic sphere.[2] A science of collective intelligence based on this model would make it possible not only to clarify the mechanisms of symbolic cognition but also to improve our collective management of knowledge and thus ultimately support human development.

The first part of this paper examines the nature of symbolic cognition, which is intrinsically social and cultural. The second part explores problems linked to knowledge

---

1 Pierre Lévy holds a Canada Research Chair in Collective Intelligence at the Department of Communication, University of Ottawa.

2 The term *noosphere* denotes the emergent sphere of human thought and the *semantic sphere* is a linguistic-mathematical model of the abstract space where this cognitive emergence takes place.

management and the maintenance of the "commons" that is human knowledge at the global level. The third part gives a broad outline of the IEML semantic sphere, a linguistic-mathematical model of the basic playing-field of collective intelligence. This semantic sphere can be seen as a huge semantic graph whose vertices and edges are variables of the same group of symmetric transformations. The fourth and last part recalls the end of the scientific revolution in the 17th century which would represent a (qualitative and quantitative) mathematization of the life of the mind and invites the reader to contemplate the unified, anthropocentric cosmos of a future civilization of collective intelligence.

Symbolic cognition

Beyond "collective stupidity"

Since I published my book on collective intelligence in 1994, I have constantly encountered the ultra-classic (and in my opinion, feeble) objection that individual humans are the ones who are intelligent, while groups, more or less organized communities, and to an even greater extent crowds, most of the times are stupid. First of all, it should be pointed out that almost all individual cognitive abilities depend on the use of symbolic (languages, systems of writing, various social institutions) or material tools (instruments for measurement, observation and calculation, vehicles and transportation networks, etc.) that the individual has not invented him/herself but which have been transmitted or taught by the pervading culture. Most of the knowledge displayed by those who maintain that intelligence is purely individual comes to them from others, via social institutions such as the family, school, or the media, and this knowledge could not have been accumulated and improved on without long intergenerational chains of transmission. Moreover, the most advanced contemporary societies depend on institutions whose main driving force is precisely collective intelligence: one could mention in particular democracy, the market, and science.

The principles of *democracy* do not guarantee that inept or corrupt leaders will never be elected, or that extremist or violent orientations can never be adopted by the majority of a population. It is merely that universal suffrage, political pluralism, the balance of powers, freedom of expression for all and the respect for human rights in general (and minorities in particular) are more conducive to civil peace and human development than dictatorships or regimes dominated by a closed group of privileged individuals.

The existence of a *free market regulated by law* will doubtless never prevent economic crises or inequalities of wealth. But the experience of history shows that planned economies in which a small number of bureaucrats determine the trends in production and fix prices are much less efficient than market economies in which all producers and consumers contribute – after a fashion and with all kinds of imbalances – to setting prices and variations in production and consumption (see Surowiecki (2004) for a recent clarification in this regard). I should make clear, in order to prevent any misunderstanding, that this perspective is obviously open to government interventions aimed at making markets more dynamic and better oriented towards human development, especially through the construction of infrastructures, through the creation of circumstances that are favourable to education and research, or through the establishment of welfare programs.

Finally, the *scientific community* is governed by principles of collective intelligence such as peer review, the reading and citing of colleagues, the reproducibility of observations, and data sharing. All these principles fail to protect against repetitive mediocrity, or against errors or "false" theories. But the collective intelligence of the scientific community is obviously preferable, in the progress of knowledge, to arguments of authority or hierarchical, dogmatic, opaque institutions invested with inquisitorial powers.

In the digital environment, the success stories of open-source software and Wikipedia are well-known illustrations of the power of collective intelligence.

Thus, the unnecessary sarcasm about collective stupidity (which is obviously always the stupidity of "others") fails to recognize everything our individual wisdom owes to tradition and what our most powerful institutions owe to our ability to think and decide together. Need one add that the adoption of collective intelligence as an essential value implies no abdication of critical thinking or individual originality? Collective intelligence, as I am defending it here, is the opposite of conformism or stifling standardization. Recognizing fully what we owe to the traditions or communities in which we participate involves precisely the moral obligation to enrich "the commons" in return by an effort toward original, relevant creation. Collective intelligence is only enriched by linking or coordinating individualities, by facilitating dialogues, not by levelling off differences or silencing dissidents. Finally – need it be recalled? – no common knowledge can be created, accumulated or transmitted without a *personal* effort to learn.

From animal collective to human collective intelligence

The first scientists to have studied collective intelligence are the ethologists, those who observe and model animal behaviour. They have taught us that, even if the cognitive abilities of individual ants or bees are quite limited, anthills or beehives, on the other hand, when considered as "wholes" or "super-organisms", are capable of solving complex problems in a coordinated fashion (Wilson, 1971; Wilson, 1975; Bonabeau and Theraulaz, 1994; Bonabeau *et al.*, 1999). The composition or aggregation of numerous simple, individual forms of behaviour can result in a subtle, complex, social behaviour that goes beyond the individuals' understanding. Collective intelligence exists not only in insect societies but in schools of fish, flocks of birds, herds of herbivores, packs of wolves or troops of monkeys, as well. In general, the fact of living in societies whose individuals communicate and cooperate provides a competitive advantage for a great number of animal species.

Humanity is a highly social species and, as such, displays properties of collective intelligence just as other social primates do. But there are several notable differences between human and animal collective intelligences.[3] The source of these differences is the innate biological ability of human individuals to *manipulate symbols*, whether these symbols be linguistic, iconic, musical, etc. We must of course distinguish between communication and language. Many mammals or birds of the same species, for example, are able to communicate amongst themselves to signal sources of food or the approach of predators. It is also clear that mammals, in particular, are skilled at communicating their emotions:

3 Henceforth I shall take "animal" to mean the non-human animal, although strictly speaking humans are obviously also animals.

aggressivity, fear, joy, desire to mate… Communication may even be sufficiently elaborate as to encompass meta-communication, for example in play activities (Bateson, 1972). But all that still does not mean that animals either manipulate symbols or have a command of language. Because they lack the linguistic ability that distinguishes humans, animals can neither ask questions, nor tell stories, nor participate in dialogues. Although a small number of ethologists believe that some animal societies have a "culture", that is, a certain ability to transmit invented or learned behaviour, it is clear that nothing equivalent to the history of technologies, of music or of political forms (for example) exists in the animal kingdom and particularly not within the same species. In contrast to the cultural (thus human) history of architecture, beavers have been building their lodges the same way for as long as beavers have existed. From the point of view that interests us here, two essential differences distinguish animal from human collective intelligence.

The first difference is that the human individual not only has remarkable problem-solving abilities, but also and above all, has a *reflexive consciousness* (Bickerton, 1995), which is afforded him by discursive thought, whether the discourse underlying his thought be internal or interactive.[4] Animal organisms are certainly also "aware" in the sense that their organisms undergo subjective experience, particularly perceptions, sensations of pleasure and pain, emotions, etc. But they do not engage in autonomous reflection on their own behaviour. They do not think discursively about what they do before, during and after their actions for the good reason that, without language,[5] they have quite simply no way of entertaining any rational thought whatsoever. They do not represent themselves in the mirror of their own discourse. We should therefore not represent human collective intelligence to ourselves, patterned on animal collective intelligence, as emerging from the interaction of unreflecting behaviours, without the room for autonomy afforded by discursive thought. No, in humans, a threshold has been crossed. Because human collective intelligence gathers, connects and makes up individual cognitive processes that are radically more complex than those of animal collective intelligence, cognitive processes that are in some way enlightened from within by discursive reason[6].

The second difference is that human collective intelligence unfolds, from one generation to another, in the course of a history or a *cultural evolution*. For example, the history of the material processes of production and processing, over the long term, shows an increase in the human species' control over its environment. Once pottery and metallurgy were invented, these processes were transmitted, improved on and added to previous acquisitions in the history of technologies. The same is also true for the media or for sign systems such as, for example, writing or currency. In general, inventions that permanently increase the power of the societies that use them are kept, whether they relate to material procedures or symbolic institutions. Unlike animal collective intelligence, human collective intelligence learns not only at the level of one generation in time, or the level of one society in space, but also at the much vaster space-time level of the human species as a whole.

---

4 We know that for Vygotsky (Vygostky, 1934), internal discourse is internalized dialogue.

5 Quite obviously, this absence should not be seen as a "fault". Animals, including their different cognitive styles, are all perfect in their own way.

6 Of course this reasoning can still be sick, perverse, often saturated with unconscious impulses, conditioned by somatic or cultural structures that transcend it… But none of that prevents reason from existing and remaining, in spite of it all , unique to humanity.

The secondary reflexivity of symbolic cognition

Some researchers[7] view the cognitive process as beginning in the biological functioning of the cell or of plant organisms, because of the autopoïetic and self-referential nature of living entities in general. But there is general agreement in thinking that "sensory" or "conscious" cognition begins only with animals equipped with a nervous system. Nervous systems interpose complex computational circuits *between* the sensory capture of information (the excitation of sensory nerves such as touch receptors, photoreceptors in the retina, taste buds, etc.) and the control over muscle movement. Heinz von Foerster (von Foerster, 1981), one of the leading thinkers in cybernetics and the theory of self-organizing systems, and without doubt one of the founders of "artificial life", has rightly said that the nervous system actually works much more on its own results than on the raw sense data received by sensory receptors. Indeed, in the most sophisticated animal organisms, there are many more neurons that receive their input from intermediary neurons than from neurons fed directly by sensory receptors. Perceptible qualities such as colors, shapes or odours are not received directly from the outside world but really and truly *calculated* from the excitation patterns of sensory receptors. There would be no color in a world without eyes, without optic nerves, without the complex biological computation machine of the brain, just as there would be no pleasure or pain without nerve impulses or the emission and reception of diverse chemical messages in the interdependent ecosystemic networks in which animal organisms participate.

The principal effect of neural computations is the *categorization* of the perceptible, at the most basic level first (pleasure or pain, salty or sweet, round or angular, blue or yellow) then at the level of the construction of objects such as prey, predators, partners, evidence of past events... And perception is always coloured by an *affect* (beginning with attraction or repulsion), whether this affect be conscious or unconscious. Based on the computed reception and production of chemical messages, emotions such as fear, aggressivity, sexual attraction and others infuse the perceptible world constructed by the sensory system with a whole gamut of subjective energies.

There aren't raw phenomena – on one side – and categories that would format the phenomena – on the other. Phenomenal experience and perceptual-affective categorization are on the contrary two facets of one single cognitive process. The phenomenal experience of animals is the subjective counterpart of the process of the categorization and affective colouring performed by their nervous systems. Emerging from the sensory-motor loop, a flow of phenomenal experience brings with it the distinction between objects, qualities, or poles of attraction and repulsion, which necessarily correspond to categories. There is no visual image – for example – without colour, brightness or a contour of some sort. In a nutshell: perceiving is categorizing. The categorization process that takes place in the sensory-motor loop – especially in the neural computation that takes up most of this loop – to my mind explains the *reflexivity* of animal cognition. The animal is "aware": that is, its phenomenal experience shines from within, illuminated by a sort of primordial clarity of the sensory: the (non-visual) light of categorizing experience with an

7 In particular, the Chilean school of biological philosophy represented by H. Maturana and F. Varela, see e.g. (Maturana and Varela, 1980).

affective tinge. This *primary reflexivity* of animal cognition is a constitutive element of the phenomenal world in general.

I shall now turn to symbolic cognition and its *secondary reflexivity*[8]. The characteristic of symbolic cognition is its ability to represent – and thus to conceive – the categories that organize experience *by means of classes of phenomena*. Classes of phenomena (sound, visual or others) representing categories are the *signifiers* and the categories themselves are the *signified*. In the primary reflexivity of the animal *sensorium*, the categories are implicit: they are incorporated into the *modus operandi* of the neural circuits. In the secondary reflexivity of the human intellect, the world of categories becomes explicit; it slips out of the neural wings to take the phenomenal stage. The activities of certain clusters of neurons (Changeux, 1985), those that distinguish between colours and stabilize the visual category *red*, for example, are strangely reflected in a class of sounds (or in a string of visual characters): "red" in English, "rouge" in French. And this class of phenomena is itself recognized by means of other patterns of neuronal excitation so that the brain becomes capable of representing its own activities through phenomenal images chosen (by culture) as instruments of self-reference. The symbol therefore depends on a two-tiered encoding of categories that involves not only (1) the nervous system *at the level of the categorization of phenomena*, but also (2) a conventional correspondence – established by the collective intelligence of the culture! – between signifiers and signifieds *at the level of the projection of categories onto phenomena*. The essence of language is this explicit representation of the categorization process in the phenomenal world. Because of human language, the world of categories becomes an object of cognition and the activity of knowing can therefore reflect itself. This circular process is what I call the secondary reflexivity of symbolic cognition. In summary, given that at the most abstract level the "observer" is a system of categorization of a flow of data, symbolic cognition is intrinsically reflexive *because it permits self-observation by the observer*, that is, the cognition of a system of categorization by itself. And this is only possible because the categorization system is projected into the data flow that feeds it.

The unity of symbolic cognition and the diversity of its cultural expressions

The categorization system of animals is wired in patterns of the neural circuits or programmed in metastable equilibria of hormone concentrations. This does not preclude a certain plasticity, as evidenced by the learning abilities of animals. But the fact remains that the categorizations performed by the nervous systems of non-human organisms are mainly genetically encoded at the species level, while the *symbolic* encoding of categories is not determined at the level of the human species but within the framework of cultural communities that institute and share symbolic systems. The *capacity* for symbolic manipulation and encoding is what has been determined once and for all at the species level: linguistic ability is innate or natural to the human being. But the codes themselves are adopted by cultures: languages, for example, are conventional and variable. Thanks to its ability to handle symbols, the human brain functions somewhat like a "universal machine"

8 Bickerton (Bickerton, 1995) provides an excellent analysis of the reflexive human consciousness based on linguistic ability.

that can interpret and carry out the instructions of a very large number of categorization systems, as evidenced in particular in the great diversity of languages, music systems, literary genres, religious rituals or technologies implemented throughout the history of our species. In order to grasp the nature of symbolic cognition well, we should never lose sight of two points. Firstly, the symbols (relationships between signifier and signified) never exist in isolation: they belong to *sets of symbols* that *form a system*: languages, writing systems, religions, political constitutions, economic rules, and so forth. Second, the users of these symbolic systems are never isolated, either: in order to exist as effective symbolic systems, languages, writing systems, religions, political constitutions or economic rules must function *at the level of communities or societies that form cultural units*.

As animal organisms, we participate in the collective intelligence of the primate societies to which human communities belong. But as carriers of *logos*, we participate in (symbolic) cultural cognitive systems that are much more complex than those of the bonobo or gorilla societies. In view of symbolic cognition, talking bipeds do not represent self-sufficient cognitive systems but rather interconnected processors that effect – with a margin of real but limited autonomy – the cultural computations of collective cognitive systems. These collective intelligences weave the cultural fabric, first, because they interface and, after a fashion, connect *several* languages and symbolic rules, and then because – by so doing – they *coordinate* the symbol-processing activities of the individual. This is how institutions, in the broadest sense of the term, are able to function.

At the origin of human cognitive reflexivity, there is a capacity for symbol manipulation that is more general than language, more elementary than music, myths, rituals and technology. In this regard, Jacques Derrida[9] referred to an inherent writing at the basis of our thought processes, a script that is in no way a transcript of speech and whose marks bear no traces of any prior presence. One can also connect this primordial writing system with the fundamental intuitions that are at the origin of Chomsky's "universal grammar" (Chomsky, 1957; Chomsky, 2000) or with the "language of thought" of the philosophical tradition,[10] but without limiting it to being merely the archetype of natural languages. This cognitive proto-writing can be considered, rather, the abstract basis of our general capacity to arrange any set of symbols on any grid whatsoever, and to carry out any reversible operations of reading and writing on these arrays of symbols. This universal abstract capacity for reading and writing is an innate *symbolic power* that was actualized by the hunter-gatherers of the oral cultures long before the official advent of script in human civilizations. We find this same cognitive power at the source of the three main types of games, semiotic, social and technical, that characterize humankind.

We are distinguished from other species primarily by our linguistic, narrative, musical and artistic skill in general. We enjoy producing, editing, and sending messages. No other species on the planet plays with signs as we do.[11]

---

9 Especially in his three books published in 1967: *La voix et le phénomène*, *L'écriture et la différence* and *De la grammatologie* (Derrida, 1967a; Derrida, 1967b, Derrida, 1967c).

10 See for example *Le discours intérieur. De Platon à Guillaume d'Occam* by Claude Panaccio (Panaccio, 1999). Note especially the notion of "verb" developed by Augustine in De La Trinidad, see: (Augustine, trad 1869), http://fr.wikisource.org/wiki/De_la_trinité_%28Augustin%29

11 See for example: (Barthes, 1967; Beaudrillard, 1972; Guattari, 1992).

Secondly, the complexity of our institutions and our social relationships far exceeds that of primate societies. Without going into detail regarding the delicate complexity of the legal or political structures of the great civilizations, suffice it to recall the importance of rituals in the vast majority of cultures and their significance in all milieus. In ritual, it is the people, their dress, their attitudes, their words or their songs, their actions performed collectively, each according to his/her role, that act as signifiers. Confucian teachings consider excellence in the practice of ritual (which includes not only exceptional ceremonies but also everyday behaviour altogether) to be one of the aims of scholarly education. Filial piety, familiarity with classical texts, elegance in writing, appropriateness of language and ease in performing rites, all derive from the same virtue of *humanity*[12].

Thirdly, if humans are a technically-oriented species, this is precisely because of their ability to treat any bit of matter as the occurrence of a signifier within a complex system of categories. The intricacy of an electronic circuit or an aircraft engine is of the same order as that of a tragedy in verse, a classical symphony or the design of some Persian carpets. Contemporary engineering excels as much as the sacred architecture of ancient Egypt or India in its ability to produce intricate structures and design systems of alternating symmetries. Isn't Daedalus, the Greek hero of applied science, also the architect of the labyrinth? Wielding a bow and arrow requires as much manual dexterity as holding an artist's paintbrush… and the same direct intuition – Zen – of the target. *È cosa mentale.* In their apologies on painting, Leonardo da Vinci or Vasari still struggled quietly within the old hierarchy between the plebeian, mechanical arts focused on things, and the noble, liberal arts, focused on signs.[13] But this inequality stems more from the social convention of a specific time and place than from the anthropological universal. The movements of the saw and the bow are similar: while one emits musical sounds, the other sculpts a three-dimensional form. The gestures of the musician or carpenter make sense in handicraft practices with long traditions, which are in turn inserted within broader cultural groupings. In every case, the point is to produce or manipulate signifiers, i.e. phenomena that are datable and addressable within the space-time continuum but that send out tangled roots at a distance into impalpable worlds of meanings.

The reciprocal envelopment of the phenomenal world and the world of the intellect

A symbol is classically composed of two parts: a signifying part ("signifier") and a signified part ("signified"). *The occurrences of the signifying parts of symbols* always appear to us in a phenomenal or sensory-motor form. I am speaking here of the *occurrences of signifiers* and not of *signifiers alone* because signifiers themselves are *classes* of phenomena rather than individual phenomena, which are dated and located in the space-time continuum. Let us think for example of words (signifiers), which only have a place in the virtual system of the language, by contrast with the pronunciations of these words (occurrences of signifiers), which truly occur in the space-time continuum. The processes

12 See for example: (Galison, 2003; Chang, 1997).

13 The issue was to show that painting is rooted in the liberal arts and the notion of rhetoric, even if it does not employ words.

of symbolic cognition always imply, ultimately, *classes* of interactions that are felt, recollected, imagined or dreamt with phenomenal appearances, because the concepts must be represented by signifiers. Phonemes of a language, musical notes, written characters, images of ritual, religious or artistic articles are such classes of phenomena. But in the actual cognitive processes, it is the *occurrences* of signifiers that are not only perceived but also produced, invented, processed or shifted by gestures, body movements, possibly by means of tools such as pens, paintbrushes, musical instruments… Since categories or ideas (being of an abstract nature) cannot be perceived, imagined or manipulated independently of their perceptible signifiers, we can say that for human cognition, the intellectual world of categories is necessarily enveloped in – but also veiled by – the phenomenal world.

I should now like to demonstrate that, still from the standpoint of human cognition, the phenomenal world is symmetrically enveloped in the world of the intellect, where relations between classes come into play. Let us note to begin with that the phenomena that we perceive, that we produce, and on which we act are generally named or labelled by us in one or several symbolic systems and particularly in languages. We *categorize* not only the beings and objects with which we come in contact, but also their relationships, the dynamics of these relationships and the rules of the games that these dynamics follow.

Now, whenever a phenomenon is named or categorized in any manner whatsoever, it can be treated as the occurrence of a signifier, that is to say, manipulated according to *conventional* syntactic, semantic and pragmatic rules that are specific to human communities. Thus a dynamic configuration of phenomena is meaningful only on a playing-field and according to *invisible* rules that derive from a symbolic order. This order is symbolic because it goes beyond the perceptible appearances of the phenomenal world. *In addition* it brings into play networks of categories invested with all sorts of affective energies according to highly variable rules of measurement and evaluation[14]. This is how our musical intelligence deals with sounds. We combine and decode the phonemes of language in this same spirit, i.e. based on meanings and relationships between meanings in practical situations. And our politeness refines the choreography of social relations according to complex signifying motifs rather than focusing on the mechanical relationships between bodies with mass. Almost all our interactions with perceptible phenomena can be conceived in the form of the recognition and transformation of *signifying configurations*.

I started out by saying that symbolic cognition always has a physical and phenomenal, sensory-motor counterpart. We write with our eyes and hands, we talk with our whole body. But we must also recognize that, symmetrically, all human work, all activities that belong in a cultural framework, including those that seem to be purely physical or material, *also* manipulate symbols. We interact symbolically with the phenomenal world and we manipulate images in order to have access to the universe of concepts.

This is why, for example, the secret dance performed by hunters in the forest and which involves the hours and the winds, weapons and gestures, animals and their tracks, also possesses a symbolic character that is projected from the depths of prehistory onto

---

14 In his *Philosophical Investigations* (Wittgenstein, 1958), Wittgenstein states that "language games" are not only linguistic phenomena but really and truly *forms of life*.

the walls of the Lascaux or Altamira caves. And the same is true for the complicated procedures followed by garage mechanics to disassemble and reassemble engines, or by seamstresses who sew, mend and embroider clothing. In every case, including what at first glance seems to be a series of utilitarian acts involving only the material world, humans interact with occurrences of signifiers, images or bodies that refer to complex arrangements of categories, to ideal models, evaluation criteria, levels of affective intensity, rules of the game, in fact a complete universe of symbols … To repeat, we nearly always treat actual bodies – including our own – *as signifying images*, regardless of the senses (hearing, sight, touch, smell…) to which these bodies appear or of the sensory-motor dynamics from which our cognitive assessment of the bodies in question arises. Human beings cannot undergo any (phenomenal) experience without making it meaningful. Is what happens real or fictitious, true or false, insignificant or important, good or bad, safe or dangerous, sad or joyful… and, for all of this, to what extent, against what background of meaning, practical expectations, or desires? The play *with* and *between* phenomenal bodies is thus inevitably projected onto a world of variables, operations and possible relationships *that is not the world* of material things in three-dimensional space, but that of conventional symbolic systems. Human symbolic cognition can be seen as a strange operator that reciprocally envelops and connects ideas and phenomena. Now this loop necessarily passes through collective intelligence, because symbolic systems are designed by the culture and because the phenomena that we experience daily are co-produced socially.

Knowledge management

The information economy

Our first meditation on collective intelligence centred on understanding its essence: symbolic cognition. I shall turn now to a second meditation, in which human collective intelligence is no longer to be merely contemplated in principle, but also improved in practice: the focus will therefore be on knowledge management.

This second aspect of collective intelligence - obviously dependent on the previous one - is particularly relevant in the context of the *knowledge society*, in which we constantly seek to improve more and more on both formal and informal practices of knowledge management. The main idea of the knowledge society is that human development in general, and economic prosperity in particular, require the intensive use of knowledge. From this standpoint, the collective ability to create, share, assimilate and apply knowledge constitutes one of the main driving forces of development. The concept of knowledge must be understood here in its broadest sense and is thus not limited to scientific or academic learning. It can refer to traditional knowledge, knowledge deriving from practice, skills of an artistic type or the refinement of aesthetic, ethical or emotional competences.

The concept of *information economy* is close to that of the knowledge society. There are several possible definitions. In its broadest sense (which I prefer), the information economy integrates all operations of production, exchange, recording, use and evaluation of information. In this very general sense the information economy is as old as the human species, or perhaps even older, as old as the biosphere. In human society, the information economy is supported and regulated by symbolic systems (Lévi-Strauss, 1950). Now this symbolic economy (Beaudrillard, 1972) is itself dependent on the media sphere: for example, our species has undergone several major reorganizations of its information economy, in particular those related in succession to writing, the intensive use of the printing press, or the electronic media (McLuhan, 1962; McLuhan, 1964; Goody, 1977; Eisenstein, 1983; Goody, 1987; Havelock, 1988). But transformations of the media sphere are only important to us in that they have made possible reorganizations of the information economy, in other words of collective intelligence. Each of the giant steps in the information economy has undoubtedly been as a result of the discovery of new procedures for exploiting the innate human capacity to manipulate symbols collaboratively. These procedures rely on the material resources of the media sphere but are not limited to these, since they also include symbolic systems and disciplines of the mind. We need therefore to distinguish between different styles of information economy, in other words several historical forms of production, storage, transmission, exchange, evaluation and application of knowledge.

When one speaks today of the information economy, the term is usually employed in a narrower sense than that which was just mentioned and which corresponds in fact to the idea of knowledge society outlined above. This narrow sense of information economy refers to *a particular historical period* that begins roughly in the mid-1960s (Richta, 1973). This period is marked by the global generalization of electronic media, by an acceleration in the rate of production and of obsolescence of knowledge, by the (still ongoing) international explosion of the university population, by a steady increase in the volume of information exchanged and stored and, consequently, by the critical role of information and knowledge managements in economic, social and cultural life.[15] Now the more the success (whatever our definition of success may be) of a community depends on its creative management of knowledge – which is the case today – and the more the ability to think together becomes crucial (Benkler, 2006). I would underline in passing that the capacity for collective intelligence can be noted at any level at all, from small workgroups to networks or institutions of all sizes to the entire human species. In this conceptual framework, there will be an ever-increasing tendency to characterize contemporary human groups based on their degree of collective intelligence, that is their skill in working together to maintain and improve a cycle of production, exchange and implementation of knowledge. This is merely a suspicion of a general trend, which should certainly not be hastily operationalized unconditionally. No one can fail to notice, in fact, that at the present time neither is there a serious unit of measurement nor are there rigorous scientific methods available to evaluate the strength of a collective intelligence. The few tentative efforts that have been made in this direction have generally gone no farther than choosing a series of indicators and measuring *quantities* (a "collective intelligence quotient"), whereas what is needed is a way to describe system dynamics, development patterns, and transformation models for quantities and values in the universe of meanings. And in the event that one could imagine having such a scientific method available, the classic distinction between the object being studied and the subject conducting the study would be extremely

15 See Porat, Marc & Rubin, Michael, *The Information Economy*. Government Printing Office,Washington DC, 1977 and (Castells, 1996; 1997; 1998).

difficult to maintain. It can never be guaranteed – for example – that the supposed "object" being studied (a human group) has not developed a cognitive dimension that completely escapes those who claim to be specialists in its measurement or evaluation. The science of collective intelligence that I aspire to cannot be other than completely open, dialogic and symmetrical (or reciprocal,[16] with the object and subject regularly exchanging roles).

Despite the above-mentioned difficulties in measurement, I postulate that there exists an intrinsic relationship between collective intelligence and the information economy, both in the broadest sense of the latter term, and the narrowest. From the most general perspective, the two terms are almost equivalent: to every form of information economy there corresponds a particular organization of the collective cognitive system. *The information economy is to human symbolic cognition what ecology is to the biosphere.* From the most limited perspective, which is also the most practical, the strength or fertility of collective intelligence becomes the principal factor for success and thus one of the most desirable "assets" of the information economy: its fundamental capital. In this latter case, collective intelligence focuses more specifically on creation, invention, discovery, innovation and learning (Lévy, 1994; Surowiecki, 2004; Tovey, 2008), that is everything that contributes to the maintenance and growth of the common knowledge capital.

I spoke above of the "successes" of human communities, without specifying the nature of the goals targeted. It is time to clarify this point. I am convinced – along with many others – that at the level of huge populations, nations, continents and our species as a whole, the most general goal, the most sustainable and the most consistent with the commons is what has come to be referred to as *human development*. Human development is a multidimensional notion that covers in particular economic prosperity, an increasing number of opportunities for everyone, public health and people's well-being, the control of ecological balances, the transmission and enrichment of the cultural heritages, peace, security, human rights, democracy, innovation in all fields (scientific, technical, artistic, administrative, social), etc. (Sen, 1999; United Nations, *Human Development Report*, 2002). All the dimensions of human development just listed should be thought of as interdependent, that is as different facets or conditions of the same process of cultural growth. Now human development, thus conceived, depends on the creation and the judicious and intensive use of knowledge, and thus ultimately, of collective intelligence.[17] According to this hypothesis, there is a *causal relationship* between the effectiveness of the collective intelligence of a community and its ability to solve problems of human development. It is my guess that, in the now-emerging global civilization, collective intelligence – or wisdom – will be explicitly recognized as the main driving force of human development and that, symmetrically, human development – improving people's lot and fulfilling their potential – will come to be perceived as the condition of possibility for the growth of collective intelligence (Lévy, 2007). Collective intelligence is the secret to strength.

---

16 On the concept of reciprocal anthropology, developed in particular by Alain Le Pichon, see the journal Transcultura (http://transcultura.jura.uni-sb.de/english/index.html).

17 A hypothesis formulated in particular in my 1994 work entitled, precisely, "L'intelligence collective" (Lévy, 1994).

The commons: reciprocal envelopment of the biosphere and the semantic sphere

Since the beginning of the 21st century, in conversations that weave the fabric of the new global public space, there has been much discussion of *the commons* (in French, "les biens communs"). This term, which is quite comprehensive, includes both public goods the consumption of which by some does not deprive others – such as sunsets and useful knowledge – and pooled resources that could suffer from overexploitation or lack of maintenance by some members of the communities considered – such as irrigation systems and public libraries.[18] This economic concept originally denoted above all the non-appropriated parts of the ecosystem surrounding human communities who engaged there in activities of direct removal (hunting, gathering, and wood-cutting for the *forest*), or of livestock breeding (for *pasturelands*). Historians of the United Kingdom often mention the "Enclosure Movement" led by lords and large landowners beginning in the 16th century, which had the principal effect of reducing the English "commons" to virtually nothing, like Balzac's "magic skin"... and giving the go-ahead to capitalism. The link between the concepts of the commons and ecosystem environment has become stronger nowadays. Aren't drinking water, breathable air, liveable climate and biodiversity all commons for which it is urgent to find suitable methods of management? And if so, fences around private property are not the only things threatening the sound management of the biospheric commons, but also the barriers that separate national territories. Now there is another commons that is as global and as indispensable to organized human life as the balance of diversity in the biosphere, and that is knowledge. Once again, to avoid misunderstandings, I shall take that to mean not only scientific learning approved by Academia, but also traditional knowledge and skills in many traditions and communities of practice.[19] Aside from their global, cross-cutting nature and their forming an infrastructure that is indispensable to social life, I would like to emphasize a third characteristic shared by these two major types of commons: they are dynamic, changing, interdependent systems composed of a multitude of autopoietic cycles and of intermeshed feedback loops. Indeed, the knowledge common to human societies forms something like a complex ecosystem. Like collective intelligence, of which it is one aspect, the knowledge community can be envisaged at different levels, from the small workgroup or personal social network to businesses, schools or universities, cities, regions, and all sorts of social media and virtual communities on the Internet to the entire species.

Considered from the standpoint of its unity, we already know that we must comprehend knowledge as an *ecosystem to be carefully maintained.* How might we clarify the general form and the internal dynamic of this ecosystem?

To begin with, interaction with the biological ecosystem is obviously not only a fact of hunter-gatherer and agricultural societies. The industrial and post-industrial economies are *also* forms of management and transformation of biospheric "nature": the only change is in the quantitative levels (much more massive) or the degree of fineness (bio-

---

18 See Elinor Ostrom and Charlotte Hess (Ostrom and Hess, 2006) p. 9.

19 On the concept of community of practice, see Etienne Wenger, (Wenger, 1998) and on the broader concept of ecologies of practice producing unique forms of knowledge that are irreducible to official science, see Isabelle Stengers (Stengers, 2003).

or nano-technological) of their transformations and removals. Having said that, one of the main differences between the biological ecosystem and the epistemic ecosystem is that the former offers us substances to eat and drink, to clothe and warm us, to build with (thus: materials, concrete…), while the latter only provides us with information or even just methods for interpreting information. But we should immediately correct this statement by observing that the majority of goods that we extract from the biospheric environment can only be removed by means of the knowledge and the technical know-how that we have regarding them. Admittedly, part of our common knowledge (let us take the example of literature or psychology) has no direct bearing on how best to exploit animal or plant species, the oceans, the soil or subsoil. But different types of knowledge are interconnected by means of the complex, interdependent network of culture, so that ultimately the knowledge ecosystem as a whole helps map out the material interactions in which we are immersed, guide our maintenance of the biological ecosystem, and model our removals of its flows and stocks.[20] The two major types of commons are thus closely interdependent. The collective capital that the biological ecosystem represents is actually defined by the epistemic ecosystem, which permits us to analyze, maintain, improve and exploit it. We do not live in the same "nature" as hunter-gatherers because we do not decipher it according to the same codes and thus interpret it differently. As for the common knowledge capital, the concept only acquires meaning in the network of physical, economic, technical and other interactions that we maintain with the biological ecosystem. Humans are in some way the central interface where the biological and epistemic ecosystems, or the biosphère and semantic sphere, are co-defined. Seen from another angle: our common knowledge capital is the cognitive medium that opens up access for us to our physical environment.

The symbiosis of capital and labour in the information economy

I understand the word "capital" based on its etymological meaning of "livestock" (in French, *cheptel*): a collection of heads (in Latin *caput, capitis*) of cattle. Capital would have been originally a domesticated, living population, capable of reproduction and lending itself to artificial selection. If the archetype of capital is the flock/herd, that of labour is thus the activity of shepherds, cowboys, or gauchos. To the tribe of breeders who lead their livestock to the best pastures, who direct their evolution by sound crossbreeding, who protect them from non-human predators who threaten their new-born, the herd in return supply their fat, meat, bones, skin, hair, milk, manure, warmth, power for transport or draught, their animal force. Capital and labour are in a relationship of interdependence: the life of one depends directly on the life of another. One might say that the herd of animals and the human tribe form a *symbiotic unit*. Thanks to their association, they are able to better survive and reproduce in their common ecological niche than they would do without each other. As for any symbiosis, domestication is a successful arrangement for both partners, not only for humans: plants and animals domesticated by humans greatly affect the biosphere with their huge populations.

---

20 Claude Lévi-Strauss has frequently drawn attention to the role played by classifications of the natural, physical environment of cultures in the development of their social, religious, and other categories. See especially (Lévi-Strauss, 1962).

Let us now substitute for the animal herd *a network of useful knowledge*, and for the tribe of livestock breeders, a community of *communicating thinkers* within the knowledge society. Let us remember that capital and labour maintain a symbiotic relationship. In other words, knowledge itself, on the one hand, and the activities of symbolic cognition engaged in by all members of the community together who "give life" to this knowledge, on the other, are two complementary aspects of the same autopoietic, self-organizing, evolving, fragile process: that of the information economy. We can thus address this process just as well in terms of capital as in terms of labour.

What does the "labour" of maintaining and developing knowledge consist of? Both knowledge and practice communities perform well-ordered, socially coordinated operations on symbols. According to the circumstances, these operations consist in raising or solving problems, strictly executing instructions or modifying ways of doing things, inventing new rules or repeating moves already played a thousand times. It matters little, moreover, whether these operations are mainly classified in the area of messages, social relations, technical reports or – most often – a mixture of all three. What matters is that the interaction between these operations of manipulating signifiers performed by people, in specific places and at specific times, comprise something like the activity of a collective intelligence. Labour that gives life to the capital of common knowledge is a process of *social cognition* that extends necessarily into a space-time area vaster than that of individual operations. But – beyond the space-time continuum – the movements of the information economy unfold above all in the invisible dimension in which we perceive meanings: in the virtual ecosystem of a semantic sphere with unfathomable limits, where discursive multiplicities exchange their affective energies and kaleidoscopes of ideas intertwine their semantic colours.

Our coordinated cognitive activities feed a *knowledge capital.* As a first approximation, a "unit" of knowledge is a recurrent circuit of symbolic energy in an abstract universe of meanings, a circuit that is not only in a relationship of interdependence with phenomenal experiences but also with other symbolic circuits… The work of collective intelligence aims primarily at the *reproduction* of these circuits of symbolic energy that intermesh in the semantic sphere. First, knowledge is received from a tradition and must be retransmitted. Then, these local symbolic circuits belonging to a vaster ecosystem must be *improved.* The ecosystem of symbolic circuits and practices that support them can be improved by controlled mutation, by selection and crossbreeding. The criteria for this additional value or strength are obviously reliant on a variety of changing conditions and contexts. But the guiding principle remains relatively simple: the living knowledge maintained, reproduced and improved on by a community must render useful information. This is in fact at the very heart of the symbiotic process: a population of talking primates will only maintain and refine the reproduction of circuits of symbolic energy in complex interaction *in the semantic sphere* if this knowledge ecosystem in turn helps in the reproduction and well-being of actual human bodies immersed *in the biosphere.* The knowledge must therefore favour maintenance of the physical ecosystem around the communities that keep them alive (agriculture, industry, development of biological ecosystems), improve the physical condition of this population (safety, health…), and satisfy its spiritual need to give sense to its life and its world (mutual trust, aesthetic or sacred ordering of existence). This is what

I spoke of above when I said that, within the perspective of the knowledge society, *human development* and collective intelligence involve each other.

Just now I brought up tribes of livestock breeders to indicate what I meant by the original, founding, archaic pact that links virtual knowledge ecosystems with human populations. Talking primates cannot survive without culture; symmetrically, the idea ecosystems that structure this culture can only reproduce in symbiosis with the desiring, suffering, mortal bodies of the social mammals that support them. And in revisiting our reflection, perhaps we should also consider that these idea ecosystems "breed" communities of talking primates in order for them to reproduce and diversify…

Reflective explicitation and sharing of knowledge

The majority of us today no longer live, like our ancestors, in a single tribe. Contemporary social life generally makes us participate in a multitude of communities each of which maintains a different cultural tradition or knowledge ecosystem. As members of a family, speakers of a language, citizens of a nation, followers of a religion, practitioners of a discipline, apprentices of an applied science, amateurs or masters of an art, associates in a company or an administration, or affiliated with a thousand networks, associations and workgroups, we participate in more than one cultural community. I consider these communities mainly in terms of their role in the activation, reproduction and modeling of knowledge ecosystems. They are "labour" communities in the information economy. So many communities, so many points of view on the global ecosystem of knowledge, so many communities, so many social voices in the great global dialogue of knowledge, so many more or less organized collective intelligences of cultivators of the semantic sphere. And for each one of them, the maintenance and operation of its knowledge capital, the management of its knowledge, thus comprises a major challenge. Hence the necessity for individuals who participate in the various communities that motivate these disparate knowledge ecosystems, to take a reflective step backward and to deal explicitly with the management of their own personal knowledge, in the intersection of the collective intelligences in which they participate.

Let us make an inventory of the intellectual assets or the *symbolic ordering* of the communities that maintain a knowledge capital. First of all this involves the set of *signifiers* manipulated by collective intelligences (images, sounds, multimodal signs, material environments, implications of the body). We must then consider the *systems* of signs, the symbolic structures – including the relationships between signifieds – that organize these signifiers (grammars, codes, classifications…). Thirdly we must add "abstract machines" (Deleuze and Guattari, 1980), procedures, the *pragmatic rules* according to which symbols (customs, skills, standards and conventions of all kinds) are activated or processed. These rules include in particular methods of measurement, evaluation and judgment that give depth to the meanings. Finally, we must consider a fourth element of the symbolic ordering: its reflexive reification, its self-referential modelling, or what knowledge management experts call its *explicitation*.

The father of contemporary knowledge management, Ikujiro Nonaka (Nonaka and Takeuchi, 1995), has proposed a cyclical model of the cognitive life of organizations. According to the SECI (Socialization, Externalization, Combination, Internalization) model, knowledge exists first in *implicit* form in individual practices. These practices are then socialized (S) and shared informally to become incorporated into organizational cultures. The critical phase of knowledge management in organizations according to Nonaka is the *transition from implicit knowledge to explicit knowledge* (E). This externalization originates in a practice of questioning and dialogue that can only flourish in an atmosphere of trust. It consists essentially in representing the greatest possible number of informal practices and the pervading culture in the form of written documents, software or databases. The explicitation of knowledge has many advantages: it makes it possible to *decontextualize* and therefore distribute and share information on a large scale, critically examine the state of knowledge and even perhaps automate their implementation. The externalization of knowledge takes the form of explicit concepts, (computing) or classifications, methodological documents, rules, algorithms or programs. Once the knowledge is formalized in concepts and rules, it can be distributed in the company's information system, combined (C) and applied – perhaps in an automated form – to the flows of data that indicate the internal state or the environment of the organization. Ultimately, the results of the phases of explicitation and combination must be *integrated* or internalized (I) by the associates in order to be implemented, tested and perhaps transformed in practice, thus giving rise to a new cycle of socialization, questioning, dialogue, formalization, recombination and so on. Knowledge of the organization is the cycle of life itself, and not one or another of its constituent phases artificially isolated.

The SECI model was developed at a time when the Internet already existed but when the Web was just coming into existence and the social media were still unknown, except for a few pioneers of virtual communities. The vision that we have today of knowledge management, calls much more for collaborative learning networks based on the use of social media than on the administration of central information systems controlled by specialists. The issue is to promote organizational cultures and technological environments that favour transparency and flexible reorganization of networks of competences and the continuous, collaborative creation of immediately usable knowledge. But the fact remains that the dialectic of the explicitation, combination and practical integration remains relevant. In addition, I hypothesize that there can be no systematic exploitation or efficient maintenance of either the knowledge capital or the symbolic ordering of a community, without some explicit modeling of its own intellectual and social functioning. Its *reflective reification* constitutes one of the focal points in a cycle of production, processing and exploiting of knowledge. The three comments below clarify the main relationships that – in my opinion – unite the symbolic orderings of collective intelligence and the models that explain them.

The first point on which I would insist here is a symbolic ordering or knowledge capital should not be confused with its explicit modelling. The map is not the territory.[21] A legal code does not encompass the living system of a nation's morals. A dictionary and grammar of English provide only a partial, stop-action view of a multifaceted, evolving language spoken by a population of speakers spread over five continents. An explicit

21 See Alfred Korzybski, *Science and sanity, An Introduction to Non-Aristotelian Systems and General Semantics* (Korzybski, 1933).

model is *less* than the symbolic ordering, which it allows to be reflected and spread. It represents only an abstraction – and, I would add, one of the possible abstractions – of this reality.

My second point is in some way complementary to the first: there is no model that does not co-produce the reality it models. The map gives rise to a territory where previously there were only experiences of movement and memories of distances covered (Latour, 1985). By its perlocutionary force (Austin, 1962), the code of laws transforms the morals of a nation. Dictionaries and grammars influence academic learning and the literacy practices of languages (Auroux, 1994). The model is a *factor* in the symbolic ordering that it makes explicit.

Thirdly, the types of technical support or media used by the externalization of knowledge profoundly determine the identities of symbolic orderings and their referents. The old hand-drawn portolan charts used by mediæval sailors, maps printed according to Mercator geometric projection, and the dynamic online mappings that combine GPS, satellite images, dizzying *zoom in* and *zoom out* on the screen of an I-phone, all structure the relationships to space and travel differently. Knowledge that is reflected and transmitted through stories told in song does not have the same flavour as when it is logically formalized in writing. And if this same knowledge is represented by a database accessible online and by computer programs that automate reasoning procedures, we are dealing with a third, still different, scenario. The *medium* of the model articulates not only the model itself, but also the collective intelligence on which the symbolic ordering is modelled.[22]

In order to reproduce, develop and increase its knowledge capital, a community must have available an explicit modelling method for the cycles of cognitive operations that it performs on data flows. It must make itself a (multimedia) image of the signifiers, systems of concepts and pragmatic rules involved in its operations and it must be able to filter, retrieve, synthesize, analyzer and comment on data that accumulate in its technical memory according to the relevant divisions of its symbolic ordering. One of the main effects of knowledge explicitation, it has been said, is to allow its "distribution" beyond the geographical and social contexts where this knowledge emerged. In short, knowledge must be reified and disseminated in order that it can become more *shareable*. Its services can thus benefit a wider community. Rather than locking it in bunkers and Balkanizing it within small, closed communities, the decompartmentalizing, exchangeability and commensurability of knowledge is obviously one of the flaunted ideals of knowledge management. A collective intelligence has a truly common knowledge capital to the exact extent that its symbolic representation is explicitly modeled or reified in a *unifying medium*.

The question of the symbolic medium

Now the Internet clearly constitutes at present *the* unifying medium regarding the *technologies* for the material communication of messages; we still do not have available a *symbolic* medium or common language that would allow us to share knowledge in a computable (to

22 The role of the media in the formation of symbolic representations will not be discussed in detail in the text. Among a huge body of academic material on this subject, I shall make do with indicating here the work of McLuhan (McLuhan, 1962; 1964) and my own (Lévy, 1990; 1994; 1997).

exploit the computing power of cyberspace) and transparent fashion and thus to sustain an information economy at the global scale, with all the benefits that would be expected regarding human development. Now we cannot, strictly speaking, talk of knowledge as a commons that is effectively exploitable by each and every one according to the respective purposes and points of views of all communities, unless we have such a symbolic medium available to us.

By comparison, the invention of currency has already made it possible to mobilize, measure and calculate *quantities of value*. What now needs to be mobilized, measured and transformed automatically are the meanings and even *circuits for processing meaning*: symbolic orderings. Still in the area of mobilization, measurement and calculation, one can also draw a parallel between the unification of geographical space-time and that of semantic space. For a long time, every major cultural area had its own mapping system and its own particular "centre of the world" (e.g. Mount Meru, Jerusalem). In addition, the concept of mapping or the *geometric* coordinate system, though already conceived of in antiquity, only began to establish itself in the practice of sailors sailing far off into the Atlantic at the time of Europe's "great discoveries", that is, at the turn of the 15th-16th centuries. The issue of geometrization is very important because by itself it is enough to permit the calculation of angles, distances and positions. It must be remembered that the geographic coordinate system in use today – of meridians and parallels – did not begin to be effectively universal until the 18th and 19th centuries, supported by printing and European imperialism (Collectif, 1980). As for the universal measure of time that makes it possible to coordinate the flights of all the aircraft in the world (the system of time zones), this was only instituted at the beginning of the 20th century, after much debate (Galison, 2003). The systems of space-time coordinates, at the same time universal (which is what makes them useful) and cultural (these are symbolic conventions, tools built with a purpose in mind), have gone hand in hand with travel, trade and planetary unification[23] throughout the last three centuries. By analogy, we can consider that the present fragmentation and relative opacity of cyberspace stems from the absence of a system of coordinates or a common semantic medium beyond the multiplicity of disciplines, languages, classification systems and the universes of discourse. If such a coordinate system could mark out a virtually infinite semantic sphere, then immediately processes of collective intelligence no matter how cross-cutting, heterogeneous and diverse they may be, could begin to be observed – to be reflected on – in the immanent mirror of cyberspace. If we had available a common semantic medium then immediately knowledge management would have a new method of explicitation that would usher it into a new phase of transcultural and "geometrical" transparency and efficiency. If a protocol for modeling symbolic orderings could establish standard exchanges of semantic metadata and makes every possible game of indexing, classification, search and circulation of value and, well beyond the simple availability of documents, we will have begun to maintain something like a real *commons* of information economy.

---

[23] Confrontational unification, but unification all the same.

The IEML Semantic Sphere

Collective intelligence augmented by digital technologies

It will be clear by now that one of the main hypotheses of my research program is that there exists a relationship of interdependence between collective intelligence and human development. Now since digital technologies offer us more and more powerful means to augment our personal and collective cognitive processes, it then becomes essential to understand precisely which technical and techno-cultural factors bring about this increase (Lévy, 2000). The theme of augmenting collective intelligence through digital networks clearly constitutes a *new field of scientific research* (Kapetanios, 2009; Nguyen Ngoc *et al.*, 2009), as shown by the abundant literature on knowledge management (Nonaka and Takeuchi, 1995; Wenger, 1998; Moray *et al.*, 2000; Dalkir, 2005), as well as the interest shown in social computing and social media using digital technology (Weinberger, 2007; Tapscott and Williams, 2007; Pascu, 2008; Shirky, 2008; Li and Bernoff, 2008). Douglas Engelbart could be considered the main founder of this new field of research. He was one of the first to understand the importance that computers would acquire in augmenting the creative abilities of both individuals and groups (Engelbart, 1962; Bardini, 2000). In the 1960s, computers were still huge, terribly expensive machines, stored in refrigerated rooms and fed with data by means of punch cards by scientists in white coats. Nevertheless, even then, Douglas Engelbart was already working on the development of collaborative devices using digital technology and with interfaces (mouse, multi-windowing, icons, hypertexting) that would become so popular twenty years later. I should like to tell a story about this. At a symposium on the topic of *Philosophy and Computing* at which he was the guest speaker, I had the privilege of discussing the topic of collective intelligence with this pioneer. A Philosophy professor who was listening to our conversation came out with the usual objection about "collective stupidity". I replied that collective intelligence was *a program of scientific and technical research* and not beatific endorsement of just any collective expression. Understood in this way as a research program, the opposite of collective intelligence was not collective stupidity but really the field of artificial intelligence (AI). The historic AI of the second half of the 20th century aimed at *simulating*, or even exceeding, the cognitive performance of an *individual* through an information processing automaton. On the other hand, the research program of collective intelligence, initiated by Douglas Engelbart and a few others, aimed at *increasing* the cognitive performance of individuals and *groups* by means of a *communication environment* full of information processing automata. After listening carefully to my answer to the Philosophy professor, Douglas Engelbart warmly approved of the approach to collective intelligence that I had just described. Certainly, research into artificial intelligence has led to interesting theoretical advances in the domain of cognitive sciences and to many very useful technological innovations. But it is clear that the technological, cultural and social evolution over the past thirty years has overwhelmingly confirmed the relevance of the alternative program, that of collective intelligence, particularly with the development of personal computing for everyone, the Internet, the Web, the social media and "augmented reality" with wireless mobiles with access to cyberspace. To mention only two important authors, Henry Jenkins, one of the best analysts of contemporary popular culture, has clearly demonstrated in his book *Convergence Culture* (Jenkins, 2006) that collective intelligence and participatory culture were the main directions in the development of communication supported by contemporary digital media. Tim O'Reilly, the inventor of the term "Web 2.0", makes the whole issue of innovation in digital communication turn explicitly around the notions of "collective intelligence" and "collective mind".[24]

Metalanguage

I myself have been involved in the development of this field of research since the end of the 1980s, through the invention of knowledge trees (Lévy and Authier, 1992), several publications (Lévy, 1987; 1990; 1992; 1994; 1995; 1997; 2000; 2002) and finally with the creation in 2002 of the first academic research centre exclusively devoted to this subject (The Canada Research Chair in Collective Intelligence at the University of Ottawa, or CRC-CI). I started from the findings that the contemporary media sphere was rapidly evolving – but non-linearly – towards a globally interconnected memory of digital data. Furthermore, a diverse, colourful community – which is tending asymptotically towards the totality of the species – feeds and exploits that memory in real time. The storage capacities of the universal, digital media sphere are growing and shall continue to grow at an accelerating pace. Finally, this media sphere is becoming accessible through an increasingly ubiquitous network that provides access to steadily-increasing, decentralized computing powers. Now none of the symbolic systems invented by humankind to date is designed to take advantage of a media sphere possessing these characteristics. And this is for the simple reason that the Web has only existed for less than a generation. So I based my research program on the technical development and exploitation of an artificial "post-web" language, specifically designed to exploit the full potential of the media sphere that we see taking shape before our eyes. This symbolic system is called IEML (Information Economy MetaLanguage).

As a first approximation, IEML is a universal meaning (or concept) encoding system whose main property is to make it possible to automate operations on meanings. And I stress that these operations *are not limited* to the automation of logical reasoning, which is the hallmark of artificial intelligence. This metalanguage is intended to be developed and used collaboratively for an optimal exploitation of the possibilities of cyberspace to augment human collective intelligence. The immediate purpose of IEML is to solve *the problem of semantic interoperability* – the "digital chaos" – arising from the multitude of natural languages, classification systems and ontologies. IEML functions as a "pivot language", a system for addressing concepts that can connect different data categorization and organization systems that would otherwise remain incompatible.

I am well aware that the very idea of a universal system for encoding meaning can evoke the worst totalitarian fantasies, or at the very least the threat of a possible impoverishment of the diversity of meanings. I would therefore recall the fact that digital sound encoding and the use of universal file formats for recording music have in no way standardized musical messages but, quite the contrary, have contributed to increasing the diversity of productions, variations, mixes, exchanges and explorations of the musical universe.

---

24 See, by Tim O'Reilly and John Battelle "Web Squared: Web 2.0 Five Years On" http://www.web2summit.com/web2009/public/schedule/detail/10194

Similarly, far from standardizing the world of visual icons, the digital encoding of images using pixels[25] has facilitated the assisted production, the automated handling, and the widespread, open creation of images of all kinds. Finally, the digital encoding of written characters is at the heart of all word processing systems and no one claims that word processing software has ever limited the freedom of writing. Thanks to an open, collaborative dictionary, a recombinable set of basic operations and a practically infinite group of transformations, IEML encoding is supposed to reflect any specific meaning as a *moment* among a vast range of processing cycles, as a *node* in a multitude of networks, or as a *figure* that only appears as such against an infinitely explorable background. This is to say that its effect will be rather to open up semantic horizons than to shut them in.

The long-term goal for IEML is to become the addressing system for a common semantic sphere in which human collective intelligence can (a) organize its memory and knowledge according to an open multitude of commensurable perspectives and (b) represent and observe its own self-organizing processes. IEML is a formal language in the sense that syntactically valid expressions can be generated and recognized by a finite automaton. The number of valid expressions with discrete meanings is finite, but enormous since there are *more* of these than there are photons in the universe, and by several orders of magnitude. Every valid expression of the metalanguage can be regarded as the address of a "point" in the IEML semantic space. Points or nodes in the semantic space are connected by two types of links: paradigmatic links, which connect expressions considered according to their aspect as *concepts*, and syntagmatic links, which connect the expressions considered according to their aspect as *statements*.

The paradigmatic links themselves belong to several families: they can be etymological, taxonomical, symmetrical and serial. *Etymological* links connect compound concepts to more elementary concepts from which the meaning of the compound concepts can be derived. Typically, the meaning of a word can come from one or several roots. For example, *hypothesis* comes from *thesis* – the notion *to posit*, in ancient Greek – and from *hypo* – the spatial intuition of *below*, in ancient Greek. *Taxonomic* links connect concepts such that some are subsets of others, as in the case of "virtue" and "generosity". The links of *symmetry* connect concepts such that some can be substituted for others to exhaust a domain, as for example the different concepts of colours, and complementary concepts, as for example "to teach" and "to learn" or "professor" and "student. Lastly, there are *serial* links, marking the relationships before/after between concepts, relationships that result from an automatic organization of concepts according to linear gradients, as for example more abstract/more concrete.

Syntagmatic links design *trees* that decompose from one degree of complexity to the next: thus, discourses into composite propositions; composite propositions into sentences; sentences into words; and words into morphemes. It goes without saying that relations between morphemes in the same word, between words in the same sentence, between sentences in the same proposition, and between propositions in the same discourse, are also represented by explicit links.

---

[25] Basically, a pixel is a set of five numbers: abscissa position, ordinate position, quantity of blue, quantity of red, quantity of green.

The IEML semantic sphere can thus be seen as an immense, closed structure of syntagmatic and paradigmatic channels that crisscross around nodes (the valid expressions of IEML), each of which is a different variable in a system of symmetric, computable transformations. One can imagine the IEML *syntax* as a fixed machine capable of computing the vast, fractally complex network of the semantic sphere. This syntax machine needs to be provided with a *dictionary* establishing a correspondence between IEML and natural languages and adjusting the details of these connections. The dictionary functions as a "program" for generating the semantic sphere (an evolving, improvable program) that actualizes the potential of the computational machine that is the syntax of the metalanguage.[26]

Each point, junction or node of the IEML semantic sphere is at the centre of a multitude of computable *transformation pathways*. Along these transformation pathways, each "step" from one junction to the next is the variable of a discrete function. Step by step, little by little, these paths connect each point to the vast totality of the other points. In the centrifugal direction, a junction-point is thus the singular origin of a transformation star that generates the whole of the sphere. In the centripetal direction, a junction-point functions as a universal vanishing point for the semantic sphere, since there is a computable transformation path that leads to it from any other point whatsoever. In short, the IEML semantic sphere is a sphere whose centre is everywhere, whose circumference is nowhere, and whose every singularity organizes an enormous semantic circuit in an original manner.

The links between two points are called channels. Each channel in the semantic sphere can be associated with a symbolic energy defined by a couple of numbers: a cardinal number (*quantity* of energy; positive or zero) and an ordinal number (*quality* of energy; negative, positive or zero). The symbolic energy associated with a channel is an *information flow*. The functions of the information economy have information flows as input and information flows as output. They use only computable transformations on the numbers and on the channels. The information economy is thus a group of computable transformations of the semantic sphere (which is itself a computable system of transformations). One could therefore model all kinds of economic functions, inspired for example by ecosystem, neural, social, psychosocial or economic models in the sense of monetary economy. One could also explore the original patterns of information circulation in the semantic sphere that are specially designed to model different forms of collective intelligence. The semantic sphere can be seen as the common, transparent, computational province of a great number of *games*, each game corresponding to a set of economic functions involving certain particular circuits.

The IEML information economy offers a computable model of symbolic cognition that can be used to simulate, represent, or observe symbolic cognition phenomena at both the personal and the collective levels and thus, ultimately, render the human collective intelligence capable of reflexivity or self-reference. Practically speaking, the effective use of the model is obviously subordinated to the existence of interpretation programs that transform data from the Web into information flows in the semantic sphere, and to

26 See the formalization work in progress: http://www.ieml.org/spip.php?article156 and http://www.ieml.org/spip.php?article152

interface functions, which transform the dynamics of information flow in the semantic sphere into interactive multimedia representations.

Usage

The IEML semantic sphere is only a philosophical idea in 2010: a simple theoretical construction of a mathematical-linguistic type. But its open-source computer implementation and its collaborative use could allow us to take on two challenges linked to collective intelligence: that of modeling symbolic cognition and that of improving knowledge management in cyberspace.

The IEML semantic sphere would first make available a scientific model of symbolic cognition whose complexity would be of the same order of magnitude as its object and which would make it possible to take into account its socio-cultural dimensions. The model of the information economy in the semantic sphere does not reduce the functioning of thought to that of neurons (or for that matter to the functioning of any material system), nor does it reduce it to a few rules of automatic reasoning in a micro-world of elementary propositions. It is nevertheless scientific inasmuch as it can be explained in terms of computable functions and can be handled automatically in a transparent, reproducible and shareable manner. Representing the complex phenomena of the life of the mind by graphs or by the traffic of whatever quantities into graphs may not be entirely new. The novelty comes, not from the representation as a network, but from the fact that the vertices and edges of the semantic sphere graph are variables in a single system of computable transformations, which renders all aspects of the graph accessible to computation.[27]

Beyond the strictly theoretical issues of the modeling of symbolic cognition, the implementation of the IEML semantic sphere would allow us to move progressively towards a better integration of the heterogeneous multitudes of knowledge traditions benefitting our collective intelligence. The information economy modeled in the IEML semantic sphere offers a kind of accounting of the knowledge flows created, exchanged and evaluated in the ecosystem circuit of human collective intelligence. Knowledge stemming from an abundance of disciplines, traditions and cultural communities that are today rather poorly interconnected on the Web could be explained and coordinated much more efficiently through the IEML semantic sphere. The crucial point is that this practical integration would not pay any price in the reduction of diversity or planning down of differences but rather would be accompanied by a highlighting and articulation of the diversity of the universes of dicourse. The practically infinite and a-centric character of the semantic sphere should reassure those who fear that, with the pretext of explaining and sharing, the encoding would impoverish the knowledge or that a few dominant points of views would impose their simplistic imperialisms.

---

27 The Giant Global Graph or "Data Web" by Tim Berners Lee obviously does not satisfy this constraint as its core addresses, URIs are opaque by design, see http:// www.w3.org/DesignIssues/Axioms.html#opaque.

The unity of nature

A scientific quest for the unity of nature

The work that led me towards a mathematic-linguistic modeling of the semantic sphere is located at the convergence of various traditions in the quest for the unity of nature. These traditions are spiritual, philosophical and scientific, but I shall focus here particularly on the scientific quest. "Matter" and "spirit", the world of perceptible bodies and that of intelligible ideas, the objects of the exact sciences and those of the human sciences clearly interact and certainly form part of the same reality. The *fact* of the unity of nature can quite easily be the object of general agreement. But the real difficulty comes from the lack of a common *scientific model* of this unity. Before Galileo and Newton, the celestial world and the terrestrial (or "sublunary") world were still thought to obey different modeling schemes. Heaven, sheltering hierarchies of angelic intelligences, was the place for perfect geometric movements, while the crudely material, sublunary world was subject to processes of generation and corruption without geometric reason. Alchemy mixed practical formulas for transforming materials with symbolic practices of spiritual transformation inherited from a distant, pre-monotheistic past. The modern revolution in the experimental sciences collected all perceptible phenomena in the same universal, infinite, three-dimensional space and reduced the essential core of scientific explanations to the mathematical formalization of *causal mechanisms* (no matter how complex, indeterminate and irreversible these mechanisms may be). It must be understood here that the only thing required to be able to speak of a "mechanism" is its description in terms of computable functions. Quarks, atoms, molecules, organisms, the biosphere, planets, stars and galaxies all form part, in principle, of the same material universe coordinated by a single space-time continuum, and the sciences that study these objects can therefore communicate with each other. What about the objects of the human sciences, for example, prices, governments, social movements, literary works or rituals? Let us begin with the most striking similarity: the objects of the human sciences belong to an environment in which *quantities* really exist, just as in the material world, as shown by the extensive use of statistics in the social sciences. But the symbolic universe in which human collective intelligence belongs has in addition two "dimensions" that are lacking in the material universe: *value* (such as the result of a moral judgment: good or bad) and *meaning*. Now neither value nor meaning can be directly *situated* in three-dimensional space, although they can be indirectly connected to the material world through our cognitive processes. Even if great philosophers such as Spinoza and Leibniz in their time thought rigorously about the unity of nature, the scientific revolution of the 17th century remained unfinished: on the technical level of mathematical modeling, nature as conceived by science, because it does not include culture, which is to say, human collective intelligence coordinated by symbolic systems, is still incomplete and fragmented.

Not only is the dialogue difficult between human sciences and "natural sciences" (as if the human universe of culture were not part of nature!) because the two types of science do not share the same system of coordinates, but the problem is made doubly difficult by the disciplinary and theoretical fragmentation of the humanities and social sciences themselves. Even if we confine ourselves to the culture sciences, we still have no *common system of cultural coordinates*. In the early 21st century, we cannot yet formalize "co-computable" mechanisms to describe the processes of symbolic cognition. Each social science, each discipline of the "humanities" has its own sphere of reference, which is not necessarily compatible with that of the others, even though one can certainly feel that there are resonances and complementarities among the fields of economics, psychology, sociology, history, linguistics, literature, etc. Is this situation of fragmentation normal, natural, fixed, and desirable, as most scholars believe today? Or is it, as I believe, a state belonging to a certain time and place in the evolution of culture and the history of science?

From the 17th to the mid-20th centuries, scholars mainly used the printed word as a means of recording and communicating. And to automate their computing they had only slow, rigid mechanical machines. Since the beginning of the 21st century, thanks to cyberspace, data memory has virtually no limits on its volume and data communication worldwide has become instantaneous. As for automatic computing, its speed, distributed power, programmable flexibility – not to mention its intuitive, interactive multimedia control – has reached heights inaccessible to the imagination of previous generations. It therefore falls to our generation and the following ones – on the strength of this increase in our cognitive abilities – to complete the construction of an immense, unique, inexhaustible and scientifically determinable nature that would *include* human collective intelligence. The de-fragmentation of humanities and social sciences is obviously an essential element in this work program.

I am starting from the principal that human collective intelligence constitutes a phenomenon of nature or a particular category of natural phenomena. Now the data on this topic are being produced more and more spontaneously and accumulated in cyberspace by human communities themselves. Nevertheless, at the beginning of the 21st century, we have no way to synthesize – and thus to observe – a dynamic image of the collective intelligence that is involved in the digital medium.

The famous cover image[28] of this publication represents very well the decentralized structure of the Internet and if one identifies the nodes, it gives an idea of the source and quantity of the information flows that are exchanged on the Web. However, such an image does not give the slightest idea of the *content* of the information exchanged, and even less so of *the relationships between the meanings* of these information flows. The statistics of the search engines can provide us with indications as to the variations in the popularity of certain words over time. But these are words in natural languages and not language-independent concepts. No search engine, no social medium currently offers us a dynamic, explorable representation of the relative distribution and interrelationship of the concepts in searches, of messages exchanged, or of documents posted on the web. Now the least one could ask of a useful, scientific representation of the collective intelligence that is involved in this cyberspace is that the relationships between meanings could be mapped.

In order to usefully examine and interpret the data produced by – and reflecting – collective intelligence, we need suitable observation instruments and practical units of measurement. The *communal* nature of these tools is a *sine qua non* for an open scientific dialogue. It is not a question of envisaging the establishment of a science without standard

---

28 See http://opte.org/maps/ for an explanation of how such images are generated.

units of measurement and without the sharing of open-source, interoperable instruments of observation. At an even more fundamental level – and precisely to ensure the consistency of the observation instruments and units of measurement specific to the study of collective intelligence – we need a universal system of coordinates (i.e. not only *common* but also sufficiently broad, deep and open as to be *all-inclusive*) on which we could "project" the phenomena of collective intelligence in all their unlimited semantic variety and with all the precision one could ask for. Ordinary three-dimensional space or the geographic system of coordinates of the earth's surface obviously does not provide a suitable model. A concept, idea or meaning can*not* be located precisely, because of the type of coordinate system that has just been mentioned: *where* are justice, the number twelve or the color red located? A concept has no space-time address. This does not, however, prevent there being well-definable relationships and operations among the concepts. In everyday life we use *natural languages* to identify ideas and their relationships. But natural languages, from the fact of their multiplicity and irregularity, do not lend themselves to computability, "geometric" projection or interoperability that are required here. This is why I propose – as a working hypothesis and for lack of a better instrument – the adoption of the IEML semantic sphere as a system of coordinates on which to project the processes of collective intelligence.

It is clear that everything we can perceive, imagine or know of the inexhaustible immensity of nature is a *product* of the openly evolving cognitive system that human collective intelligence is. Collective intelligence is in some way the active mirror of nature as we can know it. We have no access to nature that is not a reflection of this cosmic mirror. On the other hand, symmetrically, it is impossible for us to observe this mirror independently of what it reflects. As I have elaborated on at length above, the ideas or categories that organize our phenomenal experience always appear to us in the form of perceptible or imagined signifiers. And these signifiers are recorded, communicated, processed by a host of very material devices and machines that obviously play important roles in the functioning of collective intelligence. Our bodies and artefacts are immersed in a biosphere on which they depend – and beyond that, in a dizzying, ultra-complex universe of interacting masses and energies.[29] The *Ouroboros* eats its tail: the scientific or mythical-traditional representation of the cosmos from which human collective intelligence emerges is itself the product of this collective intelligence, and the former evolves in complexity when the latter grows in power. The empirical and the transcendental co-emerge and co-evolve. The metaphor of the mirror is relevant inasmuch as it is impossible to observe the shiny side of a mirror reflecting nothing: human collective intelligence is inseparable from the nature it reflects and to which it belongs. But it is misleading to imagine a nature independent of the cognitive processes that reflect it: the phenomenal side of nature (that is, the nature *that appears to us*) is, again, inseparable from the human collective intelligence that structures, observes, experiences and transforms it.

29 The Ouroboros, a snake or dragon biting its tail, is a universal symbolic representation of the autopoietic cycle of nature … or of the co-dependence of the heaven of ideas and the phenomenal earth. See http://fr.wikipedia.org/wiki/Ouroboros and http://en.wikipedia.org/wiki/Ouroboros.

Nature as a "great symbol"

It has been long been recognized that the only reality of which we have direct knowledge is that of our subjective experience lived in the present, second by second. This flow of phenomenal experience unfolds in a unified *sensorium* that interweaves our five classic "senses" and our internal cenesthetic sense. Based on this original *environment* of experience, and thanks to the social coordination of its activities augmented by the manipulation and exchange of symbols, human cognition actively produces an Earth and a Heaven: with the Earth, the practical objectivity of a material world and, with the Heaven, the existential need for a world of meanings and values. I would clarify that the great diversity of cosmologies and world-views throughout human experience has shown that not only symbolic universes (culture) are conventional and dependent on a place and period. *All* worlds, including material and non-human worlds, are socially and culturally constructed or created. For example, within the Western scholarly tradition and in the natural sciences, which have continued this tradition since the 17th century, there have been different competing theories regarding material nature. Since scientific paradigms follow one after another[30] in time, it is clear that no one of them represents any stable exteriority of material nature that would contrast with the arbitrariness and variability of cultural conventions. In other words, there is no objective, material world that is independent of the social, cultural and technical context that permit us to build and think collectively. A few generations ago, the earth was still flat and situated at the centre of the universe. Because they affect the *sensorium* and the cognitive processes, new systems of coordinates, new instruments of measurement and observation, new media, new symbolic tools for description and computing create the conditions for new scientific and practical "objectivities".

Based on its environment or its source, which is the flux of experience in the present, the unity of nature differentiates into a virtual world and an actual world. To put it simply, the *actual* world is filled with processes or entities *that have a space-time address*. Contemporary physics coordinates these addresses in Einsteinian relativistic four-dimensional space-time before, perhaps, string theory models the fundamental unified field with a space of 11, 13 or 27[31] dimensions. In the very depths of the cosmos lies an ultra-complex quantum-relativistic transformation group where dynamic exchanges of masses, energies and space-time addresses take place.

On the other hand, the virtual world contains the symbolically encoded data of personal and social memory as well as all the games of interpretation and evaluation of these data. Even if the data and their interpretations are inevitably supported by material processes and entities, their meanings and values (and this is what matters to us here) belong to the *virtual* world. Because, from the point of view of the heaven of ideas, the data are perceived as "meaning vectors": they release an inexhaustible multitude of concepts that are thought by the discursive intellect and its hermeneutic activity. Now the signifieds, general categories or classes, just like their symbolic values, *have no space-time addresses.*

30 On the concept of successive paradigms in the history of science, see the classic work of Thomas Kuhn, (Kuhn, 1962).

31 See for example (Greene, 1999)

This *virtual* world of symbolic abstractions is what I propose to address through the IEML semantic sphere.

Let me be quite clear. Once again, I do not claim that the actual and virtual worlds are separate. They are continually transformed and translated into each other and are fundamentally *interdependent,* since neither one is ever more than a useful projection or creation of the environments of personal experiences of humans who must coordinate their practical activities and their semantic processing. The virtual and the actual are not two separate substances, but rather two limit categories, two poles of the same natural reality that can only be distinguished conceptually. Body and mind are merely categories that we use to organize our experience, not solid realities existing independently of our cognitive activities. The space-time world of perceptible bodies can only appear to us because it is always organized by categories already (differences of poles, qualities, objects…) and the intellectual world that houses these categories only has meaning and consistency in reference to some sensory experience that refers to a reality of a body or between bodies. The world of material bodies and that of immaterial meanings must therefore be captured in the unity of an *informational nature* that connects them through the environment of human experience. At the subtlest pole of nature, in the highest part of the Heaven of virtuality, unfolds the inexhaustible space of intellectual essences. At the most heavily material pole of nature, at the lowest part of the Earth of actuality, extends the huge vibratory complex of mass-energy, the "unified field" of physics from which the objects and the interactions emerge that we distinguish in our phenomenal experience. The entire complexity of natural processes extends *between* these two poles, between these two limit-spaces, between the relativistic space-time of "matter-energy" and the vast fractaloid network of the semantic sphere explored by the human discursive capacity. These two spaces – these two groups of computable transformations – are not themselves the objects of sensory experience but rather, I repeat, conventional abstractions that enable coordination of the multitude of instantaneous experiences that make up the "seconds of human existence". According to this cosmology, the unity of nature thus has a *symbolic structure,* since it *connects* a sensory half that expresses material configurations and an intelligible half that expresses the structures of semantic relationships. Through multi-layer translation processes with fractal complexities, this symbolic nature organizes the correspondence between a signifying phenomenal dynamic and a signified conceptual dynamic. Human collective intelligence (the dialogue between the environments of experience) generates and connects the two halves of the natural symbol: it plunges "downward" into the pole of material interaction through its sensory experience, rooted in the body, and beyond that, in biospheric interdependence; "upward", it reaches the intelligible pole of the semantic sphere through its collective capacity to manipulate signs. Between these two poles, we must think of the *unity of nature,* including the infinitely branched networks of the circulation of information among dimensions, layers and levels of complexity.

Amplified by the cosmologies of Newton and Einstein, the Copernican revolution drove our tiny, wandering planet from the centre of the material universe. The Darwinian revolution dated and located our species on the great tree of biological evolution. Well done! The Earth that supports us is not the absolute centre of three-dimensional space. Humanity did not appear in the universe from the beginning of time and was not the goal of the biosphere's evolutionary mechanism: it forms a small, late-burgeoning, serendipitous branch of the genealogical bush of life. By broadening our horizons, the natural sciences has driven us from the centre of the universe, where traditional cultures had settled us. But the broadening of horizons and the displacement from the centre must be carefully distinguished. The first scientific revolution freed our perspectives because it is logically rigorous, because it practices mathematical modeling, because it is based on public, shareable observation data, and above all because it uses instruments of observation, memory and communication that are more powerful than those of societies from before the printing press. But I claim that it has only driven us from the centre because it has remained unfinished, limited to the material half of the world. If the scientific revolution needed to be completed to the extent of including the teeming complexity of traditions and games of the human psyche, our species – under its avatar of collective intelligence – would perhaps return to inhabit the "environment" of an integral cosmos. But it would no longer be the same closed, immobile cosmos – now gone forever – of traditional societies. A science reconciled with the unity of nature, a second scientific revolution, would present us with an open, dynamic, creative, evolving universe in which collective intelligence – the forever imperfect mistress who is nonetheless responsible for the great symbolic game – would explore the active interface between the unlimited Earth of phenomenal configurations and the boundless Heaven of conceptual constellations.

Acknowledgements

The work presented here has primarily been funded since 2002 by the Canadian Federal Government through its Canada Research Chairs program. I also received two research grants from the Social Sciences and Humanities Research Council (SSHRC) of Canada. I would like to thank Darcia Labrosse, Michel Biezunski, Steve Newcomb, Andrew Roczniak, Christian Desjardins and Samuel Szoniecky for their collaboration.

References

Augustin (1869) *De la Trinité* trad. Duchassaing et Devoille, Bar le Duc, http://fr.wikisource.org/wiki/De_la_trinit%C3%A9_%28Augustin%29

Auroux, S. (1994) *La Révolution technologique de la grammatisation*. Mardaga, Liège.

Austin, J. L. (1962) *How to Do Things With Words*, Oxford U. P.

Bardini, T. (2000) *Bootstrapping, Coevolution, and the Origins of Personal Computing*, Stanford University Press.

Barthes, R. (1967) *Le système de la mode*, Seuil, Paris.

Bateson, G. (1972) *Steps to an Ecology of Mind*, 2 vol., Chandler, New York.

Beaudrillard, J. (1972) *Pour une critique de l'économie politique du signe*, Gallimard, Paris.

Benkler, Y. (2006) *The Wealth of Networks: How Social Production Transforms Markets and Freedom*, Yale UP.

Bickerton, D. (1995) *Language and Human Behavior*, University of Washington Press, Seattle.

Bonabeau, E., Dorigo, M. and Theraulaz, G. (1999) *Swarm Intelligence: From Natural to Artificial Systems*, Oxford University Press.

Bonabeau, E. and Theraulaz, G. (1994) *Intelligence collective*, Hermès, Paris.

Castells, M. (1996, 1997, 1998) *The Information Age*, Blackwell, Cambridge MA.

Chang, A. (1997) *Histoire de la pensée chinoise*, Seuil, Paris.

Changeux, J.-P. (1985) *Neuronal Man: The Biology of Mind*, Pantheon Books, New York.

Chomsky, N. (1957) *Syntactic Structures*, Mouton, La Hague and Paris.

Chomsky, N. (2000) *New Horizons in the Study of Language and Mind. Cambridge*, UK, Cambridge University Press.

Collectif (1980) *Cartes et figures de la terre. Catalogue de l'exposition du Centre Georges Pompidou.* Centre Georges Pompidou, Paris.

Dalkir, K. (2005) *Knowledge Management in Theory and Practice*, Elsevier, New York.

Deleuze, G. and Guattari, F. (1980) *Mille Plateaux.* Minuit, Paris.

Derrida, J. (1967a) *La voix et le phénomène*, PUF, Paris.

Derrida, J. (1967b) *L'écriture et la différence*, Seuil, Paris.

Derrida, J. (1967c) *De la grammatologie*, Minuit, Paris.

Engelbart, D. (1962) *Augmenting Human Intellect*, Technical Report, Stanford Research Institute.

Eisenstein, E. (1983) *The Printing Revolution in Early Modern Europe*, Cambridge University Press.

Galison, P. (2003) *Einstein's Clocks, Poincaré's Maps*, Norton & Co., New York.

Goody, J. (1977) *The Domestication of the Savage Mind*, Cambridge UP.

Goody, J. (1987) *The Logic of Writing and the Organization of Society*, Cambridge UP.

Greene, B. (1999) *The Elegant Universe: Superstrings, Hidden Dimensions, and the Quest for the Ultimate Theory*, Vintage, London UK.

Guattari, F. (1992) *Chaosmose*, Galilée, Paris.

Havelock, E. A. (1988) *The Muse Learns to Write, Reflections on Orality and Literacy from Antiquity to the Present*, Yale UP.

Jenkins, H. (2006) *Convergence Culture: Where Old and New Media Collide*, New York University Press, New York.

Kapetanios, E. (2009) "On the Notion of Collective Intelligence: Opportunity or Challenge?", in *Inter. Journal on Organisational and Collective Intelligence (IJOCI)* Vol. 1, Number 1, Idea Group Publishing

Korzybski, A. (1933) *Science and sanity, An Introduction to Non-Aristotelian Systems and General Semantics*, International Society for General Semantics, San Francisco.

Kuhn, T. (1962) *The Structure of Scientific revolutions*, University of Chicago Press.

Latour, B. (1962) "Les vues de l'esprit, une introduction à l'anthropologie des sciences et des techniques", *Culture technique*, Paris, p. 4-30.

Lévi-Strauss, C. (1950) "Introduction à l'œuvre de Marcel Mauss", in *Marcel Mauss, Sociologie et anthropologie.* Paris, PUF.

Lévi-Strauss, C. (1962) *La pensée sauvage*, Plon, Paris/*The Savage Mind* (1966) Chicago UP.

Lévy, P. (1987) *La Machine Univers. Création, cognition et culture informatique*, La Découverte, Paris.

Lévy, P. (1990) *Les Technologies de l'intelligence. L'avenir de la pensée à l'ère informatique*, La Découverte, Paris.

Lévy, P. and Authier, M. (1992) *Les Arbres de connaissances*, La Découverte, Paris.

Lévy, P. (1992) *De la programmation considérée comme un des beaux-arts.* La Découverte, Paris.

Lévy, P. (1994) *L'Intelligence collective. Pour une anthropologie du cyberespace*, La Découverte, Paris/*Collective Intelligence: Mankind's Emerging World in Cyberspace* (1997) Perseus Books, Cambridge, Mass.

Lévy, P. (1995) *Qu'est-ce que le virtuel?*, La Découverte, Paris/*Becoming Virtual. Reality in the Digital Age* (1998), Plenum Trade, New York.

Lévy, P. (1997) *Cyberculture*, Odile Jacob, Paris/*Cyberculture* (2001) University of Minnesota Press.

Lévy, P. (2000) *World Philosophie: le marché, le cyberespace, la conscience*, Odile Jacob, Paris.

Lévy. P. (2002) *Cyberdémocratie: Essai de philosophie politique*, Odile Jacob, Paris.

Lévy, P. (2007) "Société du savoir et développement humain", in *Le Canada et la société des savoirs*, sous la direction de Imbert P., CR Université d'Ottawa: Canada, enjeux sociaux et culturels dans une société du savoir, p. 115-175.

Li, C., Bernoff, J., and Groundswell (2008) *Winning in a World Transformed by Social Technologies*, Harvard Business Press.

Maturana, H. and Varela, F. (1980) *Autopoiesis and Cognition*, Kluwer, Dordrecht.

McLuhan, M. (1962) *The Gutenberg Galaxy*, University of Toronto Press.

McLuhan, M. (1964) *Understanding Media: The Extensions of Man*, New American Library, New York.

Morey, D., Maybury, M. and Thuraisingham, B. (Ed) (2000) *Knowledge Management, Classic and Contemporary Works*, MIT Press, Cambridge Mass.

Nguyen Ngoc, T., Kowalczyk, R. and Chen, S.-M. (Eds) (2009) Computational Collective Intelligence, Semantic Web, Social Networks and Multi-agent Systems. First International Conference, ICCCI 2009, Wroclaw, Poland, Oct 2009, Proceedings. Springer, Berlin-Heidelberg-New York.

NONAKA, I. and TAKEUCHI, H. (1995) *The Knowledge-creating Company: How Japanese Companies Create the Dynamics of Innovation*, Oxford UP, New York.

OSTROM, E. and HESS, C. (Ed) *Understanding Knowledge as a Commons*, From Theory to Practice, MIT Press, Cambridge Mass, 2006.

PANACCIO, C. (1999) *Le Discours Intérieur. De Platon à Guillaume d'Occam*, Seuil, Paris.

Pascu, C. (2008) *An Empirical Analysis of the Creation, Use and Adoption of Social Computing Applications*, European Commission, Joint Research Centre and Institute for Prospective Technological Studies.

PORAT, M. and RUBIN, M. (1977) *The Information Economy*, Government Printing Office,Washington DC.

RICHTA, R. (1973) *La Civilisation au carrefour*, Seuil, Paris (First Czech edition in 1966).

SEN, A. (1999) *Development as Freedom*, New York, Anchor Books.

SHIRKY, C. (2008) *Here Comes Everybody: The Power of Organizing Without Organizations*, Penguin.

STENGERS, I. (2003) *Cosmopolitiques* Tome 1, La Découverte, Paris.

SUROWIECKI, J. (2004) *The Wisdom of the Crowds*, Random House, London.

TAPSCOTT, Don., WILLIAMS, A. D., WIKINOMICS (2007) *How Mass Collaboration Changes Everything*, Portfolio.

TOVEY, M. (Ed) (2008) *Collective Intelligence: Creating a Prosperous World at Peace*, Oakton, VA: EIN Press.

United Nations (2002) *Human Development Report 2002*, New York and Oxford, Oxford University Press.

VON FOERSTER, H. (1981) *Observing Systems: Selected Papers of Heinz von Foerster*, Intersystems Publications, Seaside, CA.

VYGOTSKY, L. (1986) *Thought and Language*, Editor Alex Kozulin, MIT Press, revised edition (Original in Russian: 1934).

WEINBERGER, D., *Everything Is Miscellaneous: The Power of the New Digital Disorder*, Henry Holt and Co., USA, 2007.

WENGER, E. (1998) *Communities of Practice: Learning, Meaning, and Identity*, Cambridge UP.

WILSON, E. O. (1971) *The Insect Societies*, Harvard University Press.

WILSON, E. O. (1975) *Sociobiology: The New Synthesis*, Harvard University Press.

WITTGENSTEIN, L. (1958) *Philosophical Investigations*, Blackwell, Oxford.

3.2 Collective Quality: How to design collective standards of knowledge?

by Gloria Origgi[1]

Knowledge is a common good. A tiny part of our knowledge of the world is generated by our own personal experience. Relying on others is one of the most fundamental ways to acquire knowledge, not only about the external world, but also about who we are, (for instance about when and where we were born). To use Mary Douglas's striking metaphor: "Our colonisation of each others' minds is the price we pay for thought" (Douglas, 1975).

The collective dimension of knowledge is acknowledged in almost every field of thought these days, from the optimistic forecasts on the power of collective intelligence made by James Surowiecki[2], to the debate on the social dimension of knowledge within recent sociology of science and social epistemology (Shapin, 2008; Goldman, 2009). Everybody seems to accept as blatant truth that, without the import of other people's beliefs, our cognitive life wouldn't be much different than that of other animals. Yet, what is surprising in this debate is that the collective dimension of knowledge has been put forward in order to argue in favor of very different conceptions of the objectivity and the standards of quality of knowledge. On the one hand, within the so-called Big-Science debate, the collective dimension of scientific work is considered the ingredient that guarantees the objectivity of that form of high-quality beliefs we name science. On the other hand, the same social dimension has been used to argue against the high-quality standards of scientific method, for a more realistic view of common knowledge (Hardin, 2009) empowered by the wisdom of the many that can overthrow the authority of the experts.

Generations of scientists have been raised in the dogma of the impersonality and collectivity of the scientific work, against a classical view inherited from the Scientific Revolution of the scientist as an isolated genius. To mention one of the most influential

1 Gloria Origgi is a faculty member of the Ecole des Hautes Etudes en Sciences Sociales. She is based at the CNRS Institute Jean Nicod in Paris.

2 See as an example, (Surowiecki, 2004).

defenses of the collective view of science, in his famous essay on *Little Science, Big Science*, which laid the foundations of contemporary scientometrics, Derek de Solla Price writes that the social nature of collaborative work in the Big Science is the only guarantee of objectivity: scientists do not base their results on their personal qualities as artists do: scientists are interchangeable because what they do is to apply a collectively shared method of investigation of nature that has nothing to do with their own personal identity. As the zoologist J. R. Baker put it: "If Mozart had not composed that immortal work of genius, the ouverture to *Le nozze di Figaro*, no one else would have done so; but if Kekulé had not lived, structural formulae and the benzene ring would not have remained forever hidden: someone else would eventually have dreamed the same dreams".[3]

Thus, according to this view, science is objective *because* is collective, it is a collective game of peers who scrutinize each other impersonally by applying a shared scientific method that is the *Norm of Quality* of our knowledge.

But, as I said, this view contrasts with a more recent view of the collective construction of knowledge, in which the standards and norms of scientific method are replaced by the rules of aggregation of lay judgments[4].

Both approaches insist on the equation: *collective* = *objective*: to achieve an objective result, that is not too biased by personal points of views, we must be many, no matter if laymen or experts. Knowledge is objective insofar as it is impersonal, disembodied, unvarnished from any singularity and subjective wisdom.

Take for example what Clay Shirky says in his last book on the power of social networks: "We are so natively good at group effort that we often factor groups out of our thinking about the world. Many jobs that we regard as the province of a single mind actually require a crowd. Michelangelo had assistants paint part of the Sistine chapel ceiling. Thomas Edison, who had over a thousand patents in his name, managed a staff of two dozens. Even writing a book, a famously solitary pursuit, involves the work of editors, publishers, and designers. Even if we exclude groups that are just labels for shared characteristics (tall people, redheads), almost everyone belongs to multiple groups based on family, friends, work, religious affiliation, on and on. The centrality of group effort to human life means that anything that changes the way groups function will have profound ramifications for everything"[5].

Thus collectivity is everything today, and knowledge seems to be a product of collective effort. Yes, but if it is so, then where do the standards of our knowledge come from? When a group is able to work out a right answer or an accurate prediction, on the basis of *what* do we judge that the answer or the prediction is the right one? Either we knew already that it was the right one, or it is just *a posteriori* verification that can guarantee the truth and the objectivity of the conclusion. In the case of science, even if it is now a truism to acknowledge the collective aspect of the scientific enterprise, the objectivity of the results doesn't come from this collective dimension, but from the reliability of the method. A hypothetic-deductive method for inferring the theorems from the axioms of a theory, an experimental or statistical method, are the fruit of a long filtering of ideas, collective or

---

3 *Cf.* (Baker, 1943, p. 36-37, quoted in Shapin, 2008, p. 9).

4 See for example (Anderson, 2008).

5 *Cf.* (Shirky, 2008, p. 16).

singular, that have distilled through centuries the "right" way. They measure reality and make predictive models of its future possible states. Science is collective because our trust in scientific method is shared almost universally: that is why the same experiment can be replicated at the antipodes of the world and the results compared. But method is not intrinsically collective.

When we come to the more debatable case of knowledge out of aggregation rules of lay judgments, the question of objectivity becomes even harder. How do we judge the reliability and truth of a result that comes from a collective aggregation of individual opinions? How do we know that our *Google* search for a certain keyword will end up displaying the "best" information available on that keyword? We know it out of personal experience: after many trials with *Google* searches, we may come up with the conclusion that the information *Google* is able to provide at the top of its results for a certain search is good enough to be believed. But we do not have independent means of granting this knowledge on the fact that it has been collectively produced.

In his provocative article on the end of scientific method, Chris Anderson simply states that we can live in a groundless world of good matches of statistic data without caring too much about method: "*Google*'s founding philosophy is that we don't know why this page is better than that one: If the statistics of incoming links say it is, that's good enough. No semantic or causal analysis is required. That's why *Google* can translate languages without actually 'knowing' them (given equal corpus data, *Google* can translate Klingon into Farsi as easily as it can translate French into German). And why it can match ads to content without any knowledge or assumptions about the ads or the content. Speaking at the O'Reilly Emerging Technology Conference this past March, Peter Norvig, *Google*'s research director, offered an update to George Box's maxim: 'All models are wrong, and increasingly you can succeed without them.'"[6]

So, collectivity in this second sense of simple aggregation of data or lay opinions, is replacing the collective enterprise of science based on the centrality and robustness of method. But the problem remains: where do the collective standards of quality come from? When I check the grammaticality of an expression by inserting it into *Google*, I trust the answer that has the largest number of results: for example, I have checked the English spelling for the word "acknowledgment" while I was hesitating between two spellings: *acknowledgment* and *aknowledgment*: given that the first form gave me 11300000 results while the second one gave only 34 300, I have opted for the first. Of course, I was right this time, but why? Is it just a matter of "epistemic luck" or do I have any other grounds for believing this result? The only ground the people have is obviously previous experience: you have used *Google* many times, you know that it is reliable as a spelling checker because you have independent ways of controlling its reliability, like the spelling checker of your own word processor, or other written authoritative sources (like a dictionary). But is this enough to ground our knowledge? And when should your independent control of the results you obtained on *Google* stop? Is the "*good enough*" epistemic strategy good enough?

---

6 *Cf.* C. Anderson, *cit.*

In the rest of this contribution, I would like to argue that in a collective world of knowledge the problem of the standards of quality remains and is even harder to address than within the classical image of science. What is the "right" quality standard for an item? What is quality, and how to apply a common standard of quality if we aggregate in a decentralized way the opinions, tastes and biases of very different people? That is a classical philosophical question that concerned philosophers such as David Hume, who writes in his famous essay *Of the Standard of Taste*:

"The great variety of Taste, as well as of opinion, which prevails in the world, is too obvious not to have fallen under every one's observation. Men of the most confined knowledge are able to remark a difference of taste in the narrow circle of their acquaintance, even where the persons have been educated under the same government, and have early imbibed the same prejudices. But those, who can enlarge their view to contemplate distant nations and remote ages, are still more surprised at the great inconsistence and contrariety. We are apt to call barbarous whatever departs widely from our own taste and apprehension: But soon find the epithet of reproach retorted on us".

Standards of quality thus change, and each human being can cultivate his or her own idea of what is good and what is bad without harmonizing it with the others. In his essay, Hume's target was aesthetic taste and its subjective dimension and how common standards can rise and stabilize: his solution was to appeal to the experts, the *connoisseurs,* those whose expertise can be a guide for the others: "It is natural for us to seek a Standard of Taste, a rule by which the various sentiments of men may be reconciled; at least, a decision, afforded, confirming one sentiment, and condemning another." In order to achieve this, he argued, human beings have to appeal to men and women with special qualities: "Strong sense, united to delicate sentiment, improved by practice, perfected by comparison, and cleared of all prejudice, can alone entitle critics to this valuable character; and the joint verdict of such, wherever they are to be found, is the true standard of taste and beauty" [...] Though men of delicate taste be rare, they are easily to be distinguished in society, by the soundness of their understanding and the superiority of their faculties above the rest of mankind. The ascendant, which they acquire, gives prevalence to that lively approbation, with which they receive any productions of genius, and renders it generally predominant. Many men, when left to themselves, have but a faint and dubious perception of beauty, who yet are capable of relishing any fine stroke, which is pointed out to them."[7]

But the appeal to the authority of wise men outside the aesthetic domain doesn't seem to fit the rhetoric of the Modern Age and the Scientific Revolution according to which the quest of knowledge has to be based on collectively controllable experimental method *rather than* the authority of the elder masters. Indeed, there are many domains outside of aesthetics in which standards of quality matter and we don't want them to be produced by the discretionary power of an authority. Quality is not just a matter of taste when we look either for standards of *epistemic* quality, that is, the quality of knowledge we may acquire, or for standards of *food* quality, that is, not only the good or bad taste of food that we ingest, but the quality of its standards of edibility. Along the same lines, *industrial production* quality control procedures cannot be the result of the appeal to an authority.

---

7 *Cf.* (Hume, 1757, § 6; 27).

This is also true of *life parameters*, like the minimum wage, that should be based on collectively agreed standards.

The need for an objective notion of quality raises many questions that I will try to tackle in the rest of this chapter:

- Is it possible to get rid of a normative notion of quality and rely only on mechanisms of aggregation of lay judgments?
- How is a collective standard of quality constructed and maintained in a culture?
- Are there "better" and "worse" systems of quality assessment?

My point is that viewing quality as an intrinsic normative notion doesn't make it less "objective". It is a normative notion based on the historical records of an item, i.e. its *reputation* in a community. "Quality" as a term has always been employed with reference to a scale of value. In philosophy the "quality" of an item is an attribute of the item that makes it fit into a certain category. The activities of categorizing items and that of ranking are thus intrinsically dependent one on the other. Cultures produce rankings of quality standards, ratings of items, because this is our way of making sense of the world outside us, of sorting things in order to make them fit into a certain category. I will claim that quality is a normative notion insofar as it is a standard that is constructed within a particular *tradition*. What is a tradition? Traditions are evaluated taxonomies and rankings that are selected and stabilized in a culture by many different "forces", for instance:

- *Institutions:* public structures whose aim is to assure the coordination and maintenance of a collectivity.
- "*Sacred values*": those values in a culture that are deeply related to its identity and are hard to question or change.
- *Functionality:* those aspects of traditions that are socially functional and help to accomplish socially coordinated tasks.
- *Problem-solving:* traditional cultural artifacts are ways of solving practical problems of information sharing and productivity.
- *Biases:* tendencies of a culture to reinforce in a particular direction a value or a position in a ranking. For example, a culture may give a special weight to literacy because of the intrinsic value that this represents in its development.

Thus, standards of quality come from the collectively evaluated corpus of knowledge and practices that we call tradition. We trust a tradition because it imposes on our way of seeing reality, a ranking, a system of evaluation that orients us in our acquisition of information.

Let me introduce an example of a cultural artifact that is maintained and sustained as a fundamental part of our tradition by many of the various forces I have mentioned above: *Writing*. Writing is a cultural technique that was introduced at the end of the 4th millennium BCE in Mesopotamia as a device for external memory. It makes possible the reorganization of intellectual life and a new way of structuring thought, neither of which are possible in oral cultures. With the introduction of writing, one part of our cognition "leaves" the brain to be distributed among external supports. The visual representation of a society's knowledge makes it possible, on one hand to reorganize knowledge in a more useful, more 'logical', way by using, for example, lists, tables, or genealogical trees, and, on the other hand to solidify knowledge from one generation to the next. What's more, the birth of "managerial" casts who oversee cultural memory, such as scribes, astrologists, and librarians, makes possible the organization of meta-memory, that is, the set of processes for accessing and recovering cultural memory. Printing, introduced to our culture at the end of the 15th century, redistributes cultural memory, changing the configuration of the "informational pyramid" in the diffusion of knowledge. Writing, among other functions, helps us to categorize our past history. But why in the modern era of printing and the contemporary era of computers and Internet is hand-writing still so reinforced in school programs? The reason is because it is stabilized by many forces: schools, "sacred values", social pressures against illiteracy, and functionality. Even though hand-writing is a very complex graphical technique that no longer has obligatory "functionality" to acquiring writing skills (typing is enough in many contexts) other forces such as "sacred values" maintain handwriting in our school programs. Our illiterate past is still too close to give up to the sacred value of writing, even if its functional role is reducing thanks to new technologies.

Here I would like to make a more general point about the role of past evaluations and preferences in filtering information. I'll start with a parallel to some famous remarks Edmund Burke wrote about the importance of traditions. Burke was suspicious of revolutions because they risked of wiping out centuries of *tradition*, that is, of patiently collected and selected values, judgements and preferences refined throughout the ages. And this process of refinement is for Burke the essence of *civilisation*, of this thick cultural lore of judgements, values and opinions that penetrates into our minds through education and socialization and constitutes the necessary background of any form of wise thought. If we do not take into account the lore of traditions, for Burke we are condemned to reinvent the wheel at each generation. Our capacity of thinking about the world and the institutions around us is much more limited without the contribution of the preferences already aggregated in the past by others. As he says:

> *"We are afraid to put men to live and trade each on his own private stock of reason; because we suspect that this stock in each man is small, and that the individuals would do better to avail themselves of the general bank and capital of nations, and of ages"*[8]

Burke was politically wrong but, in some sense, epistemically right: there is something true in his reactionary remarks, even if their application to the analysis of the French Revolution is wrong for many reasons. One reason why his claims on revolutions are unacceptable today is that obviously not all traditions are worth preserving: the institutional biases and the social pressures that make a political tradition survive may be so wrong and oriented to defend the privileges of just one social class, that it is sometimes wiser to entrust a new generation to rethink the whole institutional design of a society from scratch. But, from an epistemic point of view, he captures the intuition that it is almost impossible to think from scratch, to know from scratch, without taking into account the

8 *Cf.* (Burke, 1790).

lore of others' preferences and values as it is filtered by a culture. This is an important epistemological point that evokes a similar, epistemological idea expressed by W.V.O. Quine in a famous article on his mentor, Rudolph Carnap: "*The lore of our fathers is a fabric of sentences. In our hands it develops and changes, through more or less arbitrary and deliberate revisions and additions of our own [...] It is a pale gray lore, black with fact and white with convention. But I have found no substantial reasons for concluding that there are any quite black threads in it, or any white ones.*"[9] That is, the lore of a tradition, even of a scientific tradition, doesn't transmit just a bunch of facts from a generation to another, but a sophisticated ensemble of judgements and conventions that together shape the way facts will be extracted and classified in a given culture at a given time.

Preferences, conventions and values that others have expressed thus play a critical role in the making of collective wisdom: they shape the reputational landscape that we use to organize our own heuristics to extract information and provide a – sometimes reliable and sometimes too biased – shortcut to what is worth keeping, remembering and preserving as knowledge. The epistemological enquiry to collective wisdom I am advocating here implies that reputation and rating systems are an essential ingredient of collective processes of knowledge: their cognitive role in extracting information doesn't depend on the intrusion of social factors than are external to the epistemological process, as many have argued. Reputation is a rational criterion of information extraction, a fundamental shortcut for cumulating knowledge in processes of collective wisdom and an ineludible filter to access facts. In my view, in an information-dense environment, where sources are in constant competition to get attention and the option of the direct verification of the information is simply not available at reasonable costs, evaluation and rankings are epistemic tools and cognitive practices that provide an inevitable shortcut to information. This is especially striking in contemporary informationally-overloaded societies, but I think it is also a permanent feature of any extraction of information from a corpus of knowledge. There is no ideal knowledge that we can adjudicate without the access to previous evaluations and adjudications of others. There are no Robinson Crusoe's minds that investigate and manipulate the world in a perfect solitude.

Thus, the standards of quality of collective knowledge are produced by a weighted incorporation in the production of our singular judgments of values filtered through time. That is what gives authority to a collectively produced piece of knowledge: we trust the wisdom not only of our contemporary crowd, but also of the past crowds who contributed to the crystallization of a tradition. This doesn't mean that we are passive receivers of the authority of a tradition: traditions are *indicators* of value, they point to the *proxies*[10] that allow us to orient ourselves in a space of knowledge we do not yet master. When we enter a new domain of knowledge or a new cultural corpus we acquire the "taste" of the authorities in the domain in order to orient ourselves (the "you have to like this" effect). Who are the "good" and who are the "bad"? This is the way in which a *canon* is constructed. Then, the more we become autonomous thinkers, we challenge these traditions, participate to transform them and create new *canons*. It's a salient feature of our contemporary knowledge world, so saturated of information, that different canons bloom and fade, rising and

9 *Cf.* (Quine,1954).

10 For the notions of *indicator* and *proxies* see (Davis *et al.*, 2010).

collapsing in a much shorter period than it used to be before the advent of the decentralized society of information. That is, quality commons are structured in received traditions that are learned and amended from one generation to another.

Collective knowledge is often presented today as a form of empowerment that frees us from the deference to experts and authorities. Nevertheless, as I have tried to argue here, experts and authorities have never been so present and influent in producing knowledge as a common achievement. Even in the impersonal game of science, as Steven Shapin has recently argued: "people and their virtues have always been pertinent to the making, maintenance, transmission and authority of knowledge"[11]. And even more in the aggregation of lay judgments, we must not forget that these lay judgments are based on received views and trust in authorities and traditions that do not come out of the blue. The power of collective knowledge is thus to articulate in a new way our trust in the transmitted authoritative views with the possibility of instantaneously sharing these values with others, thus amending our cultural traditions and making them evolve more rapidly.

References

ANDERSON, C. (2008) "The End of Theory. The Data Deluge makes the Scientific Method Obsolete", *Wired* article, online at: http://www.wired.com/science/discoveries/magazine/16-07/pb_theory

BAKER, J. (1943) *The Scientific Life*, New York, Mac Millan, p. 36-37.

BURKE, E. (1790) *Reflections on the Revolution in France*, in BURKE, E., *Works*, London, 1867.

DAVIS, K. KINGSBURY, B. and ENGLE MARRY, S. (2010) "Indicators as a Technology of Global Governance", IILJ Working Paper 2010/2, New York University School of Law.

DOUGLAS, M. (1975) *Implicit Meanings*, London, Routledge and Kegan Paul.

GOLDMAN, A. (1999) *Knowledge in a Social World*, New York, Oxford University Press.

HARDIN, R. (2009) *How Do You Know?* Princeton, Princeton University Press.

HUME, D. (1757) "Of the Standard of Taste", originally published in his *Four Dissertations*.

ORIGGI, G. (2008) "Un certain regard. Pour une épistémologie de la reputation", *Rome*, Workshop on Reputation. April 19-22.

ORIGGI, G. (*in press*) "Designing Wisdom Through the Web. The Passion of Ranking" in LANDERMORE H., ELSTER J. (Eds) *Collective Wisdom*, Cambridge University Press.

QUINE, W. V. O. (1954) "Carnap and Logical Truth" reprinted in W. V. O. QUINE (1961) *The Ways of Paradox and Other Essays*, Harvard UP, Cambridge, MA.

SHAPIN, S. (2008) *The Scientific Life*, Chicago, Chicago University Press; A. A.

SHIRKY, C. (2008) *Here Comes Everybody*, New York, Penguin.

SUROWIECKI, J. (2004) *The Wisdom of Crowds*, New York, Random House.

11 *Cf.* (Shapin, 2008, p. 4).

3.3 Web Science and Reflective Practice

by Kieron O'Hara and Wendy Hall[1]

The notion of reflective practice has been influential in professional practice of all kinds, including engineering. Reflective practice suggests that knowledge and practice are linked, that improvisation based on actual practice, as opposed to the application of formalised theory, drives understanding in many applied fields. A key feature of reflective practice is a feedback loop between actions and their effects, which need to be evaluated to produce understanding. The WWW is of course a piece of socially-embedded technology, and in many ways its progress can be seen in terms of such feedback loops, as development of formalisms and applications have social effects which are evaluated to create new opportunities for innovation. However, as we shall describe in this chapter, the WWW is embedded very deeply, and innovation can create social effects on extreme scales – for example, the creation of the practice of blogging and the massive linked blogosphere, which appeared in a very short space of time. The discipline which is called Web Science is intended to address the problem of reflective practice in a space where the feedback loops which facilitate learning happen at large scales over small time periods, increasing the danger that learning will happen at the cost of large-scale social damage (or damage to the WWW itself). In this essay we will interpret the practice of Web engineering through the lens of the reflective practice concept, using the development of the Google search engine as a case study, in order to articulate the particular issues that individuate Web engineering from other types of engineering.

1 Kieron O'Hara and Wendy Hall are members of the Intelligence, Agents, Multimedia (IAM) Group at the University of Southhampton (UK). Kieron O'Hara is a Senior Research Fellow in Electronics and Computer Science and Dame Wendy Hall is Professor of Computer Science and Dean of the Faculty of Physical and Applied Sciences.

The World Wide Web has become an extraordinarily transformative technology. Claims for its significance range from hype to scepticism, but in general most agree that its capacity for supporting communication and access to documents is orders of magnitude beyond previous technologies, and it has wrought great changes not only over the Internet and ICT, but also in the offline world, affecting the media, entertainment, politics and e-government, science and research, administration and commerce. Not only that but whole new areas of activity, such as social networking and multiplayer games in persistent virtual worlds, as well as new types of crime, have flourished using its protocols. The number of users is vast and growing, yet the decentralised structure of the Web – there is no editor of content, no quality control and anyone can link to anything – has democratised communications in all sorts of ways.

Yet for all its powerful effects on societies for good and ill, the Web is remarkably under-studied and under-theorised. There seem to be three principal reasons for this. First, it is a dauntingly large and complex structure. Second, it changes very quickly, so data soon become outdated. Third, it is a curious amalgam of technologies (hardware, protocols such as HTML and HTTP, and programming environments such as JAVA and AJAX) and human activities (the Web links not only documents and data, but people as well) and so a comprehensive overview demands multi-disciplinary skills ranging from computing, law, economics, sociology, management and organisation studies, media studies, semiotics, mathematics, as well as innumerable sub-disciplines. Too often the Web is studied as an example of a particular phenomenon – a network, or a set of computer languages, or a platform for commerce – whereas taken as a whole it is so much more.

Furthermore, the Web is not an exogenous entity. As Karl Marx once said, "philosophers have only interpreted the world in various ways: the point is to change it." Surprisingly, many have studied the Web without thinking that – if they did not like what they discovered – they could influence its development in more positive directions. It is an engineered technology, and so can be altered for the better. Conversely, many engineers have tried to change, and succeeded in changing, the Web, but if those changes are uninformed by an understanding of what the wider consequences are there is an element of risk. The internal risk of breaking the Web, somehow preventing or disincentivising the links which make up the Web network, should also not be discounted.

To this end, academics are coming together to foster a new discipline of *Web Science* (Berners-Lee *et al.*, 2006; Shadbolt and Berners-Lee 2008) to develop methods and curricula to understand the Web and provide foundations for engineering methodologies so that the Web can be changed for the better. If Web Science delivers a greater understanding of the Web, current threats can be identified and addressed, current opportunities pursued, and the Web itself can be adjusted to take account of social change.

In this chapter we consider some of the methodological difficulties of assembling this transdisciplinary amalgam of analysis and synthesis, study and engineering. Analogous problems arise in many engineering disciplines, so this is hardly untrodden ground (the 20th century planners who wished to create 'cities of the future' were grappling with problems of similar scale and complexity), but we will argue that the Web poses particular problems for its engineers as a result not only of its scale, but also of the range of scales at which it can be characterised. We begin in the next section by looking at methodological considerations, while in the following section we move on to consider how Web Science must operate, mapping Tim Berners-Lee's ideas onto a plausible engineering paradigm, and determining points of tension. We will put these abstract ideas in a concrete setting by looking at a large and complex Web phenomenon, Google.

Methodologies for studying the Web

The two major problems with the development of the Web Science paradigm are the creation of a common vocabulary and a common methodology across the relevant disciplines. Methodology is the focus of this chapter, and in this section we will discuss some of the relevant background to the investigation of the Web. In the first subsection we will discuss a commonly understood model for engineering practice which is not borne out by experience. Secondly, we will discuss some important points about computer science. In the third subsection we will look at issues surrounding transdisciplinary research such as that which will be required for the Web, while finally we will introduce the important ideas behind reflective practice.

Technical rationality

One common misconception about the relationship between engineering and science is that the latter is prior to the former. The model of technical rationality assumes that, given agreement on framing a problem and on the desired ends, the job of the engineer is to apply scientific theory to achieve them; The means are determined by a scientific engagement with the proposed and agreed ends.

This is a false picture in many ways, not least historically – the growth of technology and industry in the 18th century was independent of the development of science and mathematics, and the theories that explained the important processes of the industrial revolution did not appear until decades afterwards (O'Hara, 2010, p. 127-130). Furthermore the application of science to technological and engineering problems is not as straightforward as the model suggests. The model of technical rationality relies on three distinctions which in practice are very hard to draw: between means and ends, so that a technical procedure (means) can be applied objectively to a pre-established goal; between research and practice, so that theories can develop in isolation from their application; and between knowledge how (procedural knowledge) and knowledge that (declarative knowledge), so that action can be derived from theory.

The main issues are twofold. The complexity of engineering-problem-solving means that (a) framing a problem and (b) framing a potential solution space are extremely difficult, socially embedded, affected by all sorts of practical, cultural, institutional and financial constraints, and obstructed by the constant political dialectic between competing interest groups which will arise in a free public space hosting a plurality of values. Framing the problem is plagued by extreme uncertainty, and the characterisation of the solution space characterised by conflict.

Complex engineering problems are more often than not in effect unique, not fitting easily under generalisations, meaning that the application of abstract theory is non-trivial. Meanwhile, no problem can be considered in isolation, and any potential solution will have unintended consequences elsewhere. Neatness is an unusual property of problems;

as Russell Ackoff, one of the founders of operational research, argued. "Managers are not confronted with problems that are independent of each other, but with dynamic situations that consist of complex systems of changing problems that interact with each other. I call such situations *messes*" (Ackoff, 1979, p. 99).

Computer science

With respect to the Web, one would naturally expect to consult the discipline of computer science which like Web Science amalgamates analysis and synthesis, investigation and engineering. After all, the Web is a piece of computing technology defined by protocols and formalisms such as URIs, HTTP and HTML. However, it should be borne in mind that where the Web is concerned what counts is their *use*, not their form. URIs are a naming convention, providing strings of characters to identify resources that are the targets of hyperlinks – but that they are *identifiers* depends on people identifying things with them. HTML, the Hypertext Markup Language, would not be of general interest unless real people in the offline world marked up documents with hyperlinks. The structures and expressive resources of these protocols are essential, but equally essential to the Web is the fact that they are used. The use of these systems cannot be described by the systems themselves, and does not fall within the purview of computer science.

Computer scientists do not have the disciplinary expertise to explain why the protocols have been used as they have, or what effects additional protocols would make. The tools of computer science, such as formal verification techniques for example, can only give a partial picture. Deriving behaviour from specifications is hard enough, but in the case of the Web the specifications of formalisms tend to be developed painstakingly by international committees under the aegis of the World Wide Web Consortium (W3C – http://www.w3.org/) A rough specification of a requirement (e.g. 'a language for expressing ontologies') will be hard to map onto a formalism expressive enough for the task (e.g. OWL – the Web Ontology Language).

In the case of OWL, its developers were canny enough to understand that users would vary in their demands, and so developed it as a series of three sub-languages, ranging from OWL Full, very expressive but non-computable and without constraints to preserve consistency, to OWL Lite, computable and constrained but relatively inexpressive (McGuinness and van Harmelen, 2004). The existence of these various sublanguages shows awareness that the development of, in this case, an ontology depends not only on formal properties, but also on the informal purposes and institutional constraints on developers. Will the ontology be made up of flat hierarchies, be used a lot or demand regular editing and maintenance by non-experts? In that case, OWL Lite is the sensible option. If, on the other hand, the knowledge to be modelled is highly complex, well-understood and represents a hard-won global consensus, or if the modellers need the full expressiveness of the underlying knowledge representation language RDF (Resource Description Framework), then OWL Full is indicated. The point is that these actual operational details of the OWL language are social, economic and organisational – they are not computer science issues. The distinctions between the three sublanguages were created because of the perceived heterogeneity of ontology developers' demands, but beyond this recognition the computer scientist *qua* computer scientist has little to say. It may be, for instance, that

OWL Full remains unused because too expressive and providing too little constraint – that would not be the fault of its developers, who could not be expected to anticipate in detail the demands on it.

To take another example which we will explore in more detail later in the chapter, Google's success derives largely from the brilliance of its PageRank link analysis algorithm (Page *et al.*, 1999), which determines the relative importance of each element in a linked network. This is a very impressive piece of work, and there is much about the algorithm for the computer scientist to get her teeth into. However, this only accounts in part for its success. First of all, the computer scientist can describe the recursive definition of 'importance' that emerges from PageRank (an important page is linked to by lots of important pages), but the congruence between that operational definition and the requirements of Google's users is an extra fact. To give just one extraordinary example, in 2007 a man who had been missing for some years turned up at a police station in the United Kingdom in an apparently amnesiac state. His disappearance had actually been an insurance scam, which was uncovered by a member of the public Googling him to find a picture of him taken after his disappearance with his wife (Weaver, 2007). Nothing about PageRank tells you, however closely you look at the brilliant details of its weightings or damping factors, that it can be used for solving missing persons cases or insurance frauds in under a fifth of a second.

Secondly, PageRank has to function not in the neat world of a formal definition, but in the real world with hostile elements who will try to subvert the algorithm to ensure their own unimportant content receives a high ranking. This Google-spoofing (as it is called) demands a whole layer of adjustments to the algorithm, in an arms race between Google and the spoofers. Third, it is not only the adjustments to the algorithm that count, but also the extent to which they can be kept secret, so organisational issues such as security also become important. Relevance, context, security – these are not from the computing vocabulary, and require a wider disciplinary focus.

Transdisciplinarity

Individual disciplines work via the notion of abstraction. In computing, for example (Colburn, 2004, p. 322-325), important conceptual tools include data abstraction, procedural abstraction and language abstraction (i.e. the ascent from assembly language to higher-level languages). Abstraction entails universalisation and idealisation of diverse objects and relations but there is a serious question as to how far idealised theories and models can address or explain complex concrete problems – this is what theories of technical rationality kept bumping up against (Schön, 1983). There is a mismatch between academic knowledge production and the knowledge needed for solving embedded social problems, between knowledge supply and knowledge demand.

The need to weave approaches together to address complex, dynamic real world environments (or to borrow a term from Husserl, the life-world, *Lebenswelt*, the world that subjects may experience together) has led to the idea of *transdisciplinarity*. Transdisciplinarity is characterised by (i) the need for a radical integration of the disciplinary work involved, including linking abstract and case-specific knowledge, (ii) the involvement of a large number of stakeholders in problem definition and solution specification, (iii) transgression of disciplinary paradigms, and (iv) a focus on problem-solving rather than theorising which (v) promotes what is perceived to be the common good (Hirsch Hadorn *et al.*, 2008). Methods need to be tailored for highly complex problems and the difficulties of integrating the various problem-solving approaches.

To this end, transdisciplinary research demands three layers of knowledge. *Systems knowledge* describes and models empirical processes, mapping onto the scientific knowledge assumed in the model of technical rationality, but it must be supplemented by *target knowledge* about the needs and interests of the practitioners and stakeholders in a problem, and *transformation knowledge* about the technical, social, legal, cultural and other means of acting to transform the existing situation and to create new technologies, structures or practices (Hoffman-Riem *et al.* 2008, p. 4-5). Transdisciplinary research is a complex amalgam of activities, but some such perspective is required in order to bring diverse disciplines together to work in a complex *Lebenswelt* in whose progress many people have an interest.

Reflective practice

The criticisms of technical rationality and the requirements of transdisciplinarity each point towards the need for more knowledge, different in kind as well as quantity, about an engineering problem. Because of the *sui generis* nature of many engineering problems – this certainly applies to the Web of course – much of this knowledge, which is typically procedural rather than declarative, must be derived in practice, often in response to unforeseen challenges perceived during a project itself. This has led to the development of a theory of engineering practice called *reflective practice* (Schön, 1983).

In this methodology, the problem as initially set is not fixed in stone, as the practitioner must change her perceptions and strategy in response to uncertainty, instability and unique features of a problem. She proceeds experimentally, but not, as in the scientific context, using the logic of confirmation. Rather the logic is of affirmation; the aim is not to raise hypotheses to falsify them (as with Karl Popper's falsification logic of scientific discovery) but to create and discover new solutions that need be neither unique nor optimal. Controlled, reversible experiments are out of the question, and so each experiment that the engineer tries must as far as possible be sensitive to the needs of the context, and take into account elements of the target knowledge and transformation knowledge – the systems knowledge cannot be tested in isolation.

The method of reflective practice is an answer to the requirements of Web Science, to improve the Web in its relation to the offline world. Indeed, given the complexity of the problem space, it will be essential to develop engineering methods that use the insights of reflective practice, dynamically and recursively reconfiguring the problem specification as more knowledge is gained during the design and engineering processes themselves. In the next section, we will discuss this cyclical view of Web Science in more detail.

Web Science: reflectively engineering the Web

Engineering the Web has been described by Berners-Lee as requiring sensitivity to both technical and social concerns, as shown in Figure 1. The designer has an idea for an innovation and develops protocols, formalisms, software and hardware to realise his vision or idea, which may or may not be formally or precisely specified. However, no digital system lives in a vacuum, and its use will depend on a number of assumptions that the designer makes about the social context. If those assumptions are incorrect, then the system, however technically adequate for its task, will not perform that task. Note also that the designer cannot specify every single aspect of the system's behaviour; at some point the assumptions about social context will have to carry some functional weight. The designer's assumptions about context are an essential part of the design. For instance, email systems have been developed in response to ideas about how the connectivity of the Internet could be used as a communications network. SMTP was developed on the basis of assumptions about what people would want the communications system to carry, about organisational context and about the motives of senders (specifically that messages would be sent in good faith by a homogeneous community all of whose members would be concerned with a group of problems determined by an organisational context, so messages would be relevant to the receiver, generated in response to a genuine requirement, with a transparent meaning).

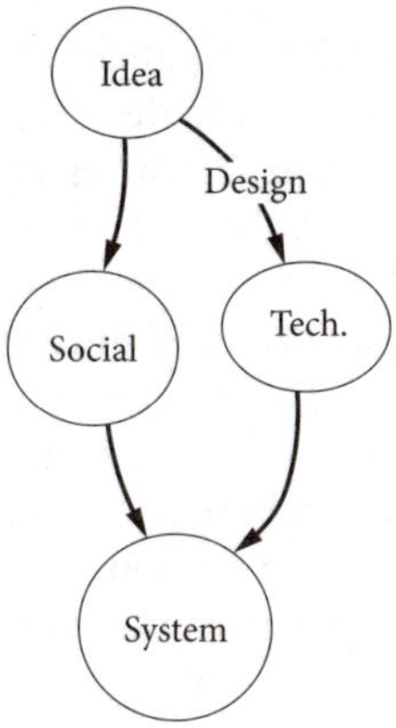

Figure 1 The Web design process (all figures in this chapter from Berners-Lee, 2007).

This is a general view of engineering. However, the Web has particular issues associated with it, because the system does not live in a vacuum any more than the design. Figure 2 shows what happens as the system is let loose in the online environment. The Web itself has many hundreds of millions of users, and billions of pages and connections, so any system can result in emergent phenomena undreamt of by the original designers, whose social assumptions can hardly be expected to be accurate in the general case. Figure 2 shows the idea of Figure 1 being implemented with some technical work and a set of social assumptions, to produce a micro-level adjustment to the Web environment, but if enough users take up a system, there will be a marked and noticeable change in macro-level perceptions. It may be that older patterns of behaviour change, or that they are supplemented by new behaviours, or that new users swell the online community (for example, consider the growth of the blogosphere, and how this has changed not only the Web, but also the media, journalism, politics, commerce and social interaction). However that may be, the end result is in effect a new Web understood at the macro level, as a result of micro-level engineering.

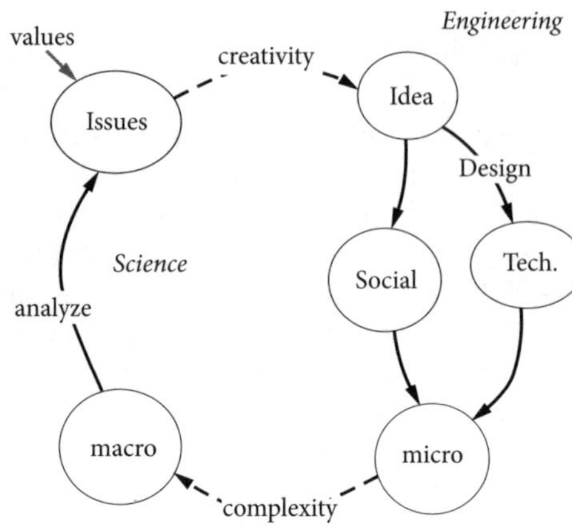

Figure 2 The Web and the world.

Those macro-level effects will need to be analysed in order to understand their wider implications, and this process of analysis will throw up important issues which will resonate either positively or negatively with the values and needs of the embedding offline societies. Unresolved issues will pique the interest of entrepreneurial designers, whose creative efforts will lead to new ideas which then have to be designed and engineered on the basis of a new set of social assumptions and the cycle begins again.

To continue the example of SMTP, the invention of email was predicated on a set of social assumptions that at implementation time were realistic (the system was largely used within the scientific community whose main concern is usually with the sharing of knowledge), but when it became a macro phenomenon, used not only by members of the target community for unforeseen types of communication but by people not in the target community at all – and ultimately the number of non-academic users has dwarfed the number of academics – the unintended consequences of a free and simple communication method became clear. Problems such as spam and phishing began to emerge. These were seen at best as nuisances, at worst as torts or frauds which raised important issues. New technical solutions, such as spam filtering, were now needed to solve the problems created by the emergent phenomena associated with the new technology. Social changes also accompanied the technology. Emails leave a semi-permanent record so it became harder for companies to hide their internal decision-making (an important factor in the prosecution of Enron, whose preserved emails now constitute a fascinating data source for those interested in corporate communication), while divorce lawyers now regularly sequester years-worth of communications from errant spouses. These developments have demanded adjustments in the law, corporate best practice, and our intuitive understanding of privacy which themselves raised more issues, and so the cycle continues.

To be sure, the Web is not the only artefact whose unintended consequences can be vast; a badly-placed bridge in a city can affect all sorts of economic relationships. Nevertheless, the disparity in scale between the micro engineering and the emergent macro phenomena is especially characteristic of the Web, whose sheer size not only spreads problems across the globe, but also creates new problems by taking design solutions out of their intended social context.

So far, Berners-Lee's characterisation of Web Science (2007) as a cyclical conversation between scientists and engineers, users and techies, fits neatly into Schön's (1983) ideas about reflective practice, while arguments parallel to Schön's show that engineering the Web cannot be a matter of technical rationality. However, his framework can also be used to point up the singularities of the Web as a piece of designed technology which demand its intensive study as a first order object as envisaged by the Web Science programme.

One point which must be made immediately is that most professional disciplines receive much of their developmental impetus from real-world requirements that are not always immediately obvious to the eye. For example, medicine has as its nominal goal wellness, but as Foucault and others have argued the meaning of 'wellness' has evolved, sometimes quite rapidly, and has had a range of interpretations from being able to live a life with minimal personal aid, to the elimination of pain, to approaching an ideal of the healthy body which itself might be determined by the medical profession or by society. However, quite aside from the nominal goal, a great deal of medical expertise is actually concerned with the problem of litigation – how to avoid it, and how, if it strikes, to construct a rationale for and defence of one's actions.

Web Science currently lacks this behind-the-scenes motivation. Its nominal goal is to improve the Web. This cannot be left solely to market structures and the straightforward profit motive; the Web is an arena for amazing innovation, but not all the innovations have been benign (and certainly their effects are rarely accurately predicted even by the startup companies that promote them). Yet there is little agreement on what constitutes an improvement of the Web. A sterner constraint – imagine a class-action suit in the US courts brought by victims of spam – would dramatically focus the effort. Without such an unwelcome focus, Web Science must include not only a debate about how to connect technical and social developments empirically and conceptually, but also a parallel debate conducted with diverse stakeholders about normative requirements.

This is a minor problem compared to the way that Web Science problems are necessarily framed and addressed. If we consider the zone of time in which an action may make a difference, what Schön calls the *action-present* (1983, p. 62), which depends on the pace of activity and the boundaries of potential action, we find it is both tiny and vast, depending on point of view. The cycles of Web development are measured in years. Blogging, for instance, took a number of years to develop from small beginnings, and then 'suddenly' took off at the beginning of the century. 'Suddenly' in this case is still a matter of years from, say, the appearance of the first blogging tools and guides and the first major political issues influenced by bloggers in 2001 and 2002, to the exponential growth characteristic of the years after 2004. But what counts is the timescale of an effective intervention. The phenomenal growth of the blogosphere was predicted by very few (as Tom Wolfe quipped, one by one Marshall McLuhan's wackiest predictions come true), and its specific effects on, say, political discourse or the offline media were anticipated by even fewer. The timescale is certainly large enough for technical development, but the social context evolves alongside the technical as well as driving it. What seems imperative in year 0 of a research project may be completely out of date by year 3 when a product appears.

So, for example, Twitter proved its political worth for many as an important conduit for news about spreading protests about the conduct of the 2009 Iranian Presidential Election, trumping traditional media outlets which were slow to feature the story. However, the downside of Twitter was also revealed at the same time when the story spread and the useful messages from inside Iran were lost in a tsunami of well-meant but pointless messages from America in support of the protesters. Furthermore, as well as describing, measuring and discussing the individual phenomena of an episode such as this, it must also be seen in the context of wider arguments, e.g. about the deleterious effect of an always-on media (Rosenberg and Feldman, 2008).

New types of online behaviour become very popular very quickly. At the time of writing (2009), Facebook and Twitter dominate thinking about cutting-edge large-scale Web phenomena, but by, say, 2014 it is quite likely that the landscape will be very different and the giants of five years previously will be hopelessly out of date. Datasets for large-scale modelling are extremely important to alleviate this issue, and some projects, such as the EU project Tagora (http://www.tagora-project.eu/), have begun to explore these spaces retrospectively. Such analyses are clearly ways forward, but as each new star application comes along, new users (possibly responding to different incentives) will arrive with it, rendering old assumptions void. Not only that, but a five or six year development and growth cycle will take many of the most enthusiastic users from adolescence to adulthood with all the attitudinal changes that implies. In short, the scale of the phenomena means that what seems a relatively long action-present for Web Science is in reality very curtailed. By the time data are gathered, models created and simulations run, the opportunity to influence events may already be past.

As noted earlier, it is characteristic of large-scale engineering that controlled experiments are impractical because their effects cannot be restricted or reversed. The relation between change and understanding is different in reflective practice precisely because of this constraint. The requirement to understand is subservient to the requirement to change for the better (unlike in disinterested research, where understanding is an autonomous goal), but the ever-present danger is that an experiment makes a permanent change for the worse. An extra and unusual issue with respect to controlled experiments on the Web is the variance of scale between the experimental setup and the outcome. Any experimental change will be of relatively small scale – a new type of software, a new type of communications protocol. The consequences *relative to the intention of the innovation* can be described and studied in small-scale experiments in the lab, or with a small set of pioneer users. Such intentions are usually focused on the experience of a single user or a single organisation. The problem, of course, is that few if any of the massive global consequences of Web technologies are of this tractable type, because they affect very large groups of people and organisations, so that even the benign or positive consequences at the scale of the Web as a whole are unintended.

Engineering using reflective practice inevitably involves trade-offs between the consequences of an artefact in relation to its intention, and the full set of consequences both intended and unintended (Schön, 1983, p. 153ff.), but the Web is an especially difficult case because the consequences in relation to the intention of the engineered development are relatively small-scale and detectable fairly quickly, while the unintended consequences, good or bad, emerge years later at a scale far beyond the control of a single person or corporation.

One way of expressing this mismatch is to look at three levels of analysis in the evaluation of design. First, the design specification includes a normative element against which it can be evaluated ("the artefact should do X"). Second, any design detail has to be evaluated against and be consistent with previous design decisions. Third, the designer must be sensitive to any new problems that arrive during deployment. As Figure 2 shows, the third level brings in phenomena at the macro scale which may take years to manifest themselves. So distributed and decentralised is the Web that even the second level is likely to be beyond the individual design team's capacity for understanding.

Another illuminating way of looking at it is through the lens of the transdisciplinarity framework as discussed above (Hirsch Hadorn *et al.*, 2008), which postulates three kinds of knowledge, of systems, targets and transformations. The Web engineer is possessed of the systems knowledge of the artefact being constructed (the node labelled 'tech' in Figure 1 and Figure 2). In those diagrams, the target knowledge (about stakeholders) corresponds most closely to the nodes marked 'social'. Hence the systems and target knowledge are, from the point of view of the individual engineering project, tractable. However, the transformation knowledge, of all the various systems relevant to implementation and use, is key, and yet is once more out of reach of the immediate designer. If we consider Figure 2, the transformation knowledge is relevant to the 'macro' phenomena on the left hand side, connected to the technical parts of Berners-Lee's diagram by a dotted line denoting the uncertainty of the connection between micro to macro phenomena.

To conclude this section, Web Science looks very like an example of reflective practice in engineering. However, the largeness of scale of the problem definition, the speed of the development cycle in comparison to the action-present and the massive disparity between the designer's understanding and the breadth of the relevant phenomena are extremely problematic. The relative absence of real-world datasets for investigation adds to the difficulties, but the retrospective nature of empirical research may mean that such investigations are anyway of only limited value relative to new systems. The need for a clear and deep integration of disciplines under the Web Science banner is evident, and is the focus of the Web Science Trust (http://webscience.org/), a research organisation dedicated to the development of methods and curricula for the analysis and engineering of the Web.

Case study: Google

How do these issues pan out in a real-world situation? Earlier in this chapter we discussed academic attempts to understand Google's PageRank algorithm as an example of the essential transdisciplinary complexity of Web-based issues. Let us now consider this company in a little more detail to see how complexity stands in the way of understanding, and how Google, in particular its founders Lawrence Page and Sergey Brin, have finessed the engineering problem (Battelle, 2005; Stross, 2008). Google's general experience fits in rather neatly into Berners-Lee's Web Science lifecycle, as he himself has pointed out (Berners-Lee, 2007, and see Figure 3).

Here the initial issue that began the cycle was the growth of the Web and the difficulty the search engines of the mid-1990s had in scaling with it (their search methods were usually designed with a particular size of Web in mind, and typically would become inadequate as the Web grew beyond that scale). The initial idea was for an index together with a support vector machine, a type of multi-dimensional classification learning method. The PageRank algorithm, which uses a linear algebra method of computing the Eigenvectors

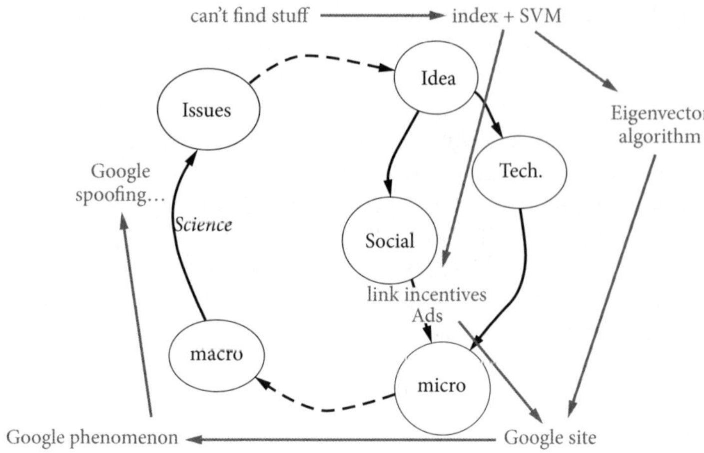

Figure 3 The growth of Google seen in Web Science terms.

of matrices representing the link structure of the Web, was the technical solution chosen, and this was expected to work in a social situation where Web authors were pleased to link to other sites they thought relevant. The result was the Google search engine, which of course grew to extraordinary size and influence.

However, new issues soon appeared. So powerful was Google that others found it began to pay to spoof PageRank, usually by creating lots of fake pages all linking to a particular page. As mentioned above, Google then had to take measures, both in terms of their algorithm and their corporate practices, to counter that. As we can see, Google's evolution fits neatly onto the schema of (Berners-Lee, 2007). In the terms of transdisciplinarity, Google's problem has often been with the transformation knowledge – the requirements of the law and intellectual property in a variety of jurisdictions is often harder than the engineering (still less was Google prepared for the controversy surrounding its move into China). Yet the issues raised by Google go beyond Berners-Lee's brief diagrammatic analysis.

The company prides itself on its array of geeky talent, but it would be a mistake to assume that this is sufficient. In particular, Google relies upon the prevailing ideology of openness on the Web. If the Web becomes transformed by a preponderance of pay-per-view sites, subscriptions, walled gardens or proprietary software (Zittrain, 2008), then Google's approach will be less powerful. The rise of Facebook in recent years, with its revivification of the walled garden model, has been perhaps one of Google's major challenges (Stross, 2008, p. 21-46).

The ideological demand for openness has some surprising effects. For instance, it precludes Google's charging for search except in certain circumstances (in its early years it provided search as a commodity for Yahoo!). That means that it must differentiate itself on quality, which has enabled it to retain its market lead (Pollock, 2008). Pollock presents an economic model of search in which users expect high quality search and refuse to pay, while search engines have the problem of finding a business model. Advertising is the key of course and advertisers want as many users as possible. In his model, this leads to a tendency towards a concentration of the market, or even monopoly (consistent with the history of the search industry so far). Monopoly does not necessarily mean that the public welfare is compromised – as long as search quality is retained, the antitrust issues may not be serious. Nevertheless accurate models are needed in order to provide predictive power, to anticipate the conditions under which concentration of the market might lead to a decline in welfare, and to inform technical and regulatory approaches to the Web.

Surprisingly, Page and Brin seem to have come late to the realisation that advertising was a potential basis for a business model (Stross, 2008, p. 3ff). The beauty of search for an advertiser is that the user's search terms indicate exactly what he is interested in at exactly the point at which the advert is served up, and this certainly unanticipated consequence of the Google model has completely subverted the advertising and media industries, to the point that mainstream media are losing money and some of their more expensive functions (quality drama, quality journalism) are being undermined for lack of a viable business model of their own.

Commitment to quality has always been a key factor for Google. One of Page and Brin's insights was that quality demands the ability to scale up with the Web, and they have invested heavily in hardware. The result is immense growth in the data storage industry (Google has been influential more widely by its promotion of the paradigm of cloud computing, which centralises storage in giant data warehouses). Here again a development which of itself creates interesting problems in the abstract (what methods for search and retrieval will work over these giant repositories?) ramifies in all sorts of unexpected ways. Cloud computing creates enormous issues for privacy preservation, for example, or the legal jurisdiction under which one's data storage falls, but the issues go beyond software or even organisational and legal structures, to hardware. A large data centre consumes about as much energy as an aluminium smelter, and policies to address climate change will certainly have implications for the continued growth of the industry.

The point of this case study is to show how quickly an idea on the Web ramifies into other areas of importance. It is certainly not intended to be critical of Google (which is a defender of the Web and has an interest in preserving its ideology of openness), but merely to show how an idea produced an ideology, new types of cheating (not by Google – we refer to Google spoofing) an unexpected business model, a monopoly, disrupted two enormous industries and will have a tangible effect on the planet's climate in centuries to come. To predict the effects of Google's business decisions will require deep analysis integrated along transdisciplinary lines. Google is not the only example of course (see (Berners-Lee, 2007) for more).

Conclusions

To conclude, the cyclic structure of Web Science is an example of the 'conversation with a problem' of the reflective practitioner. Technical knowledge (technical rationality, or systems knowledge) will not be sufficient to create a Web that serves humankind, though it *will* be sufficient to build a widget that makes the Web more dangerous, less open or less connected. The imperative for Web Science, therefore, is to try to bring together relevant expertise, to discover not only systems knowledge but also wider knowledge about stakeholders and the social context, to develop theories that evaluate designs on wider as well as more narrow criteria, and to involve as wide a range of people as possible in expressing and framing the problems it is intended to solve.

References

ACKOFF, R. (1979) "The future of operational research is past", *Journal of the Operational Research Society*, 30(2), p. 93-104.

BATTELLE, J. (2005) *The Search: How Google and its Rivals Rewrote the Rules of Business and Transformed Our Culture*, London, Nicholas Brealey Publishing.

BERNERS-LEE, T. (2007) *Looking Back, Looking Forward: The Process of Designing Things in a Very Large Space*, inaugural lecture, University of Southampton, http://www.w3.org/2007/Talks/0314-soton-tbl/#(1)

BERNERS-LEE, T., HALL, W., HENDLER, J. A., O'HARA, K., SHADBOLT, N. and WEITZNER, D. J. (2006) "A framework for Web Science", *Foundations and Trends in Web Science*, 1(1), p. 1-130.

COLBURN, T. (2004) "Methodology of computer science", in FLORIDI L. (Ed), *The Blackwell Guide to the Philosophy of Computing and Information*, Malden MA: Blackwell Publishing, p. 318-326.

HIRCH HADORN, G., BIBER-KLEMM, S., GROSSENBACHER-MANSUY, W., HOFFMAN-RIEM, H., JOYE, D., POHL, C., WIESMANN, U. and ZEMP, E. (2008) "The emergence of transdisciplinarity as a form of research", in HIRCH HADORN, G., HOFFMAN-RIEM, H., BIBER-KLEMM, S., GROSSENBACHER-MANSUY, W., JOYE, D., POHL, C., WIESMANN, U. and ZEMP, E. (Eds), *Handbook of Transdisciplinary Research*, Berlin, Springer, 2008, p. 19-39.

HOFFMAN-RIEM, H., BIBER-KLEMM, S., GROSSENBACHER-MANSUY, W., HIRCH HADORN, G., JOYE, D., POHL, C., WIESMANN, U. and ZEMP, E. (2008) "Idea of the handbook", in HIRCH HADORN, G., HOFFMAN-RIEM, H., BIBER-KLEMM, S., GROSSENBACHER-MANSUY, W., JOYE, D., POHL, C., WIESMANN, U. and ZEMP, E. (Eds), *Handbook of Transdisciplinary Research*, Berlin, Springer, 2008, p. 3-17.

MCGUINNESS, D. L. and VAN HARMELEN, F. (2004) *OWL Web Ontology Language Overview*, http://www.w3.org/TR/owl-features/

O'HARA, K. (2010) *The Enlightenment: A Beginner's Guide*, Oxford, Oneworld.

PAGE, L., BRIN, S., MOTWANI, R. and WINOGRAD, T. (1999) *The PageRank Citation Ranking: Bringing Order to the Web*, Stanford InfoLab Technical Report, http://ilpubs.stanford.edu:8090/422/1/1999-66.pdf

POLLOCK, R. (2008) *Is Google the Next Microsoft? Competition, Welfare and Regulation in Internet Search*, http://www.rufuspollock.org/economics/papers/search_engines.pdf

ROSENBERG, H. and FELDMAN C. S. (2008) *No Time To Think: The Menace of Media Speed and the 24-Hour News Cycle*, New York, Continuum.

SCHÖN, D. A. (1983) *The Reflective Practitioner: How Professionals Think In Action*, London, Maurice Temple Smith.

SHADBOLT, N. and BERNERS-LEE, T. (2008) "Web Science emerges", *Scientific American*, October 2008, p. 60-65.

STROSS, R. (2008) *Planet Google: How One Company Is Transforming Our Lives*, New York: Free Press.

WEAVER, M. (2007) "Woman found canoeist photo via Google", *The Guardian*, 6th December 2007.

ZITTRAIN, J. (2008) *The Future of the Internet: And How to Stop It*, London, Allen Lane.

3.4 Science Blogs in Research and Popularization of Science: Why, how and for whom?

by Antoine Blanchard[1]

As the Internet emerged as an efficient channel for sending information and fostering collaborations on a global scale, this unanticipated phenomenon paved the way for a new era of science, namely e-science or digital scholarship (Borgman, 2007). Massive data repositories moved online, academic publications (preprints and articles alike) became searchable across disciplinary boundaries, collaborations grew larger. But the Internet is now developing into so-called web 2.0, where active participation is replacing passive broadcasting: every user can become their own media maker and share videos, images or text. To date, the most popular form for the latter are blogs (short for web-logs). The blog format was originally used for online diaries but has rapidly evolved into a versatile publication and conversation tool.

This shift is also being embraced by scientists, on a limited, albeit growing scale. First isolated, then grouped in communities, science bloggers (I use the term to include professional scientists as well as students, journalists, science amateurs, science museums, concerned groups…) have already demonstrated the potential to influence how research is done, results are communicated and the public is reached. Among the many topics that blogs discuss, I will focus here on science and the academic life, thus defining a type of "science blogging" that is effectively turning digital scholarship into conversational scholarship (Gregg, 2006). But in concrete terms, how is that different? And where might this evolution be leading us? I try to shed some light on the matter in the following pages.

---

1 Antoine Blanchard holds a Master's degree in Sociology of Science from the University Louis-Pasteur, Strasbourg. He is currently President of the Association C@fetiers des Sciences and blogs under the pseudonym of Enro (http://www.enroweb.com/blogsciences/).

Why blog?

In 2008, an informal survey originating from the Scienceblogs community[2] was circulated among blogs, prompting for everyone's personal motivations for science blogging. Dozens of English and French-language science bloggers answered the survey on their blogs. Their answers provide a broad view of the commonly avowed reasons why people blog on the subject of science. Three main reasons stood out.

First, science bloggers find it rewarding to build an online identity, primarily aiming to convey their expertise, passion, and willingness to engage in debate. It affords them recognition (for instance when they post comments on others' blogs). It also influences the chance of their blogs being indexed by search engines and holds the attraction of possibly gaining some virtual fame or notoriety. Internet visibility may even transfer into the "real world" as the credibility of blogging as a means of exchange grows. There are now many examples of science bloggers who have been solicited by more traditional forms of media to comment on latest news or invited to address live audiences. Online presence used to take the form of personal websites that at best included a standard template for email responses. This form now falls behind these easier to use and more interactive blogging tools. Once present on the web, a blogger can start to build his or her communication network, combining the use of blogs with a vast array of other social networking tools that can be more or less tailored towards the communication, distribution and/or organization of information according to his personal tastes. (As just one purely hypothetical example to illustrate the diversity of uses; these days a researcher might make professional use of a microblogging tool[3] to keep colleagues updated on daily activities and organize meet-ups at international conferences; a social bibliographic management tool[4] to store and share scholarly articles; a visual media archive of cutting edge educational material[5]; a collaborative wiki to organize the laboratory's knowledge[6]). In general, bloggers themselves see this gamut of tools as an online version of the familiar picture of informal networking over a coffee or in conference hallways, albeit on a larger scale and with more diverse forms of engagement. That is to say, they are generally well aware that science blogging is an *additional* tool for the inquisitive mind, not a miraculous way to short cut the often mundane, daily grind of the structured, methodical component that demands will, resolution, patience, and is essential to all good research practice.

Second, science bloggers strive to make up for what they perceive as a lack of openness or transparency in research practice. Traditionally, science is portrayed as a purely methodic progress towards enlightenment, through logic and continuous discussion. Instead, science bloggers try to show other facets of what it is like to work in a laboratory and to report on the dynamic process of research in action, as seen from the front line, rather than as inferred from the publication narrative or with the hindsight of historical perspective. By creating an open forum of peers, they insert science blogging into the larger movement of open science, which also includes initiatives such as open

---

2 http://scienceblogs.com
3 e.g. http://twitter.com/
4 e.g. http://www.citeulike.org
5 e.g. http://www.scivee.tv
6 e.g. http://openwetware.org

access (article repositories or open access journals), open laboratory notebooks and open research data. Whilst the many goals and impacts of open science go beyond the scope of this paper, it is worth noting that science blogs solve one problem that was raised by the famous physicist Richard Feynman in his Nobel lecture in 1965[7]: "We have a habit in writing articles published in scientific journals to make the work as finished as possible, to cover all the tracks, to not worry about the blind alleys or to describe how you had the wrong idea first, and so on. So there isn't any place to publish, in a dignified manner, what you actually did in order to get to do the work". On science blogs, such content can find its right place.

Third, reaching out to the public at large is a high priority among science bloggers who, echoing sentiments recently expressed in print by Michel Claessens of the European Commission Research Directorate (Claessens, 2009), often find unsatisfactory the way or the frequency at which science is covered in the mainstream media. Many science bloggers see themselves as public educators, trying to raise the level of discussions e.g. on climate change or evolution (Goldstein, 2009). In France, for example, Sylvestre Huet, a science journalist with a regular short column in the newspaper *Libération* used his blog to expand coverage on the controversy over the causes of climate change and the role of the sun. This campaign, echoed by several other blogs, was sparked by a debate at the French Académie des Sciences. Researchers engaging with the wider public meet a clear demand: according to an Eurobarometer survey published in December 2007[8], "the majority of the EU27 population prefer that scientists (52%), rather than journalists (14%) present scientific information".

Other motives for science blogging that were much less explicit in the survey answers but are also important considerations include editorial freedom and adequate personal knowledge management. Bloggers feel the web spaces they create allow them to reclaim the power that they think the community has ceded to the journal editors and research institutions. They decide on their editorial policy, they set their own rules and beyond mere writing, they devise a recommendation tool that points to scientific articles, blogs (the so-called "blogroll" in the left or right column) or any other online or offline material. As opposed to the overarching goal of preserving scientific record, the blogger can even decide to contribute content of a more ephemeral nature to the so-called "web stream" and invent new forms of expression, as is the case in the development of microblogging tools such as Friendfeed[9] or Twitter[10]. Several of the science bloggers who responded to the survey on science blogger's motivation described using their blogs to "clarify ideas" and felt that by writing (and doing the research necessary to write a thought-out piece for their blog), they learned the topic better. Additionally, once they had blogged about a topic, they were able to more easily re-locate and re-use that information they had acquired, than if they had just bookmarked a page or saved a copy.

---

7 http://nobelprize.org/nobel_prizes/physics/laureates/1965/feynman-lecture.html

8 http://ec.europa.eu/public_opinion/archives/ebs/ebs_282_en.pdf

9 http://friendfeed.com

10 http://twitter.com

After this introduction to the generalities of science blogging, I now want to dive into the two main genres that make it up: blogging for research and blogging for science popularization.

Blogging for research

By "blogging for research" I refer to bloggers who are recording their work for themselves and other scientists. The process is often misleadingly viewed as synonymous with "documenting research on blogs", i.e. a way to report on someone's progress in their reflections or experiments, combining thoughts on the process along with hard data generated by the experiments. However, as the "hardcore details" are often missing, one would be led to consider this documentation either imperfect or implicit. Instead, it appears that we should not mistake the map for the territory; as blogger Jill Walker wrote back in 2003[11]: "Blogs aren't about documentation, they're about doing, thinking and discussing. And they're about catching fleeting thoughts and making them explicit" (Walker, 2006). Some research bloggers post their new ideas or their responses to published articles when they think these are important but do not need a traditional publication. Likewise, science bloggers prefer to post responses to research articles if sending a letter to the editor of the journal is simply too cumbersome. Many broad-based, pre-blogging era, research publications such as *Nature*, *Science*, *Cell*, etc have, since 2006 or so, tacked online blogging sections onto their own Internet presence. Among journals created post-blogging era, (in particular those specializing in areas of interdisciplinary interest) many have recognized the advantages of integrating the blogging process at more fundamental levels of their structural organization. Science in the making is brought to a new dimension as science blogs foster collaboration, interdisciplinary dialogue and dissemination at an unprecedented level and with unprecedented speed.

Of course not all of the new forms of rapid communication network between professional research communities come under the explicit name of a *blog*, but the principles are similar even if the precise goals are not always the same: recently for instance, among the publicly accessible discussion threads of a mailing list used within a specialist research community, a long series of fairly heated but productive exchanges about a newly commercialized research resource took place over a single week[12]. The discussion's original participants included students, post docs and group leaders from across the globe. Once underway even Nobel prize-winners Paul Nurse and Tim Hunt joined in. This particular debate brought to more general attention a long-simmering issue, which although of no importance to the broader public and of little interest to the publishing world, is ultimately a detail of paramount importance to the scientific edifice.

The first point of blogging, as I have argued above, is often to become visible and be read, or at least noticed. In a scientific world where people judge each other based on their merits and actual know how, demonstrating one's expertise by blogging can constitute a first step of being recognized as someone worth listening to. If a blogger is able to elicit

11 http://jilltxt.net/?p=184

12 http://lists.sanger.ac.uk/pipermail/pombelist/2009-October/000613.html

other people's comments and various online interactions, he can move seamlessly to networking and the construction of social credibility. As a matter of fact, bloggers increasingly use these backchannels to connect with like-minded or complementary people and carry on research projects that would have seemed impossible or painful before (Patil, 2009a). In the spirit of web 2.0, such collaborations are highly volatile and most often disaggregate as soon as they reach a conclusion. Some researchers also noted that blogs, either open or under restricted access, can serve as useful collaboration tools within research groups, allowing them to keep up with the progress of a project and collectively build a memory that comprises not only the original ideas (blog posts) but also the comments and discussions that they sparked[13].

Bloggers writing "fieldnotes in public" (Wakeford, 2008) benefit most from the readers' comments, which add to their own perspective and can improve their research. The example of philosopher John S. Wilkins is meaningful: his blog[14] on the philosophy of biology attracts a wide readership interested in topics ranging from the concept of species to the institution of academia or the creationism movement. In 2006, he ran a series of 5 blog posts about ongoing research on the concepts and causes of microbial species' existence[15]. His posts received 7 comments, a modest figure albeit quite significant in these early days of science blogging. The comments however linked to further resources and fueled a discussion that spanned quite an interdisciplinary spectrum, encompassing philosophy and microbiology alike. In his 2007 article published after some significant revision, in the academic journal *Studies in History and Philosophy of the Life Sciences*, Wilkins acknowledged his "blog readers for comment and discussion".

I have certainly benefited myself from the broad disciplines represented in the average audience of a science blog. For instance, as I posted some thoughts on how to portray "science in the making" rather than "science made", I encouraged my readers to comment and discuss abundantly[16]. Some comments asked for clarifications and pointed out unclear passages, while other suggested new examples and research for consideration. One responder noted rightly that my reflection was written from the point of view of a sociologist of science and he constructively pointed me to related works in the field of museology. I broadened my research accordingly and felt this helped me make a more general and robust point.

We call "crowdsourcing" this process of deferring to readers or online contacts (the more the better) for answering a question, recommending resources or contributing a piece of their expertise. Be the question straightforward ("Is this scientific article that was just published any good?") or more complex ("What should I read that is most relevant to what I do?"), the combination of so many answers provides "human filtered information" that is less prone to subjective bias. This can be highly useful in an era where one individual cannot hope to accumulate first-hand knowledge (tailored to one's needs or centers of interest) of every subject direct from its authoritative source. Websites such as

13 http://precedings.nature.com/documents/39/version/1
14 http://evolvingthoughts.net
15 http://scienceblogs.com/evolvingthoughts/2006/06/on_microbial_species.php
16 http://www.enroweb.com/blogsciences/index.php?2008/05/31/261

Postgenomic[17], Nature.com Blogs[18] or Streamosphere[19] tap into this collective mind to dig up the most popular science news or science articles as reported in the blogosphere. They can be more finely customized and Patil and Siegel (2009b) argue that they are often wiser than the traditional way of post-filtering the literature, based on expert recommendations (typically "editor's picks" of some top journals), reading habits and keyword searches.

Blogging for popularization of science

Participating at a panel discussion on science blogging, the blogger with pseudonym *Grrl Scientist* reported on the statistical makeup of her readership[20]: "two-thirds of my visits come from .edu or .gov sites. Primarily from educational institutions of some sort. Some people from the Senate read my blog, a lot of publishers read my blog, some lawyers… a few people who admit they're on Wall Street. I get people [aged] from 15 to people who are grandparents. I also have a classroom that reads my blog on a regular basis." Fellow blogger Brian Switek said that he makes an assessment of who is reading his blog by the types of comments he receives. For him, these are evenly divided between people who are studying science and those who are there to understand science, or those who want a "spoon-fed explanation" of the details.

The diversity of the audience is definitely a strong argument for the utility of science blogs. Whereas science museums and festivals reach a well-identified public, anyone with an Internet access can find themselves reading a science blog after a Google search or while musing on the Internet, in addition to its core readers who have established habits in the science blogosphere. Blogs are therefore a natural venue for popularizing science and engaging with the public at large. The engagement can take as many forms as the blogs themselves, ranging from comments on science news as reported by mainstream media to descriptions of, or comments on, pieces of current research from the very laboratory of the blogger. Others like to write educational pieces (e.g. a primer on the theory of evolution[21] or the purpose of the Large Hadron Collider[22]) or to highlight the science in a recent movie[23] or court decision[24]. History of science is often part of the mix, along with a tone that can be strident, sober, chatty, friendly… (Goldstein, 2009).

The content of science blogs can be very close to what one would expect from a science journalist. First, because some bloggers are actually professional journalists. Second, because less and less coverage remains the unique property of the journalists, as Bora Zivkovic a long time science blogger, argued in March 2008[25]: breaking news, be it the death of a famous scientist or the detection of water by Mars Rover, does not need much

---

17 http://www.postgenomic.com
18 http://blogs.nature.com
19 http://streamosphere.nature.com
20 http://www.underthemicroscope.com/index.php?option=com_content&task=view&id=87&Itemid=54
21 e.g. http://scienceblogs.com/clock/2006/12/from_genes_to_species_a_primer.php
22 e.g. http://science-for-everyone.over-blog.com/article-22572578.html
23 e.g. http://blog.scienceandentertainmentexchange.org
24 e.g. http://buchanan.blogs.nytimes.com/2007/05/16/the-prosecutors-fallacy
25 http://scienceblogs.com/clock/2009/03/defining_the_journalism_vs_blo.php

accompanying detail and can spark the interest of many people on its path to the mainstream media; the transmission by blogging of anticipated reports, such as the scheduled publication of a breakthrough paper or discussions at a scientific conference also increases the reach of such information to people in the right specialist community; news analysis is like second nature to bloggers with the right knowledge and the ability to find and parse through sources traditionally regarded as authoritative, such as academic or industrial scientists; even investigative reporting is taken on by the community of bloggers. For example, blogging recently blew up a case of plagiarism in dinosaur paleontology[26] and played an important role in the positive outcome of the case of the Bulgarian health care workers sentenced to death in Libya for allegedly infecting hundreds of children with the HIV virus[27].

However, there are two main differences between science blogs and the traditional way of reporting and popularizing science. The first difference is that blogs allow conversations to develop, from one blog to another but also with the readers leaving comments—while public outreach used to be unidirectional. On science blogs, people actively engage with the issues at hand, ask questions, express disapproval, while the blogger does not merely pontificate but is confronted with real-life issues that span communities. The second difference is that blogs are essentially open to anyone with an interest for a topic. They give a chance to amateur naturalists, patients investigating their disease, scientifically literate and lay people to join the conversation on an equal footing. Rather than top-down communication which effectively creates a boundary between the expert and the "non-expert", blogs offer a blend of voices and views on topics that are often more complex and less one-sided that scientists would like to believe – a point that people editing and building the online encyclopedia Wikipedia are making more obvious everyday.

Since the inception of its short life, the Internet rapidly developed a tradition of discussion forums, where people sharing a primary interest also engage in conversations. However, while undoubtedly useful, these tend to recreate the traditional boundaries between communities and levels of expertise, and they are difficult for "newbies" to enter. In contrast, blogs offer a more appealing narrative combined with a less intimidating, informal style that sparks broad discussion and makes it easier for the initially passive, or the transiently interested reader, to feel concerned enough to respond.

Blogging without boundaries

For the purpose of clarity, I have distinguished between two main objectives of science blogging even though an inherent characteristic of blogs is that they blur the categories that one tries to impose onto them. For example, the genre of science blogging often combines writings about the context as well as the content of research, making fuzzier the distinction between science *as a process* and science *as a set of results*. To illustrate my point I draw again from Jill Walker's experience in blogging about literary studies, (Walker, 2006): "What I ended up writing was a hybrid genre, my favorite kind of research blog: the blog

---

26 http://scienceblogs.com/tetrapodzoology/2008/01/aetosaurs_and_whistle-blowing.php
27 http://www.nature.com/news/2006/060925/full/news060925-2.html

that both discusses the content of research, the ideas themselves, and that also discusses the process and experience of researching." These distinctions are important ones for me as I feel that traditional science communication is too often the communication of a corpus of scientific knowledge and too rarely the portrayal of a human endeavor. Reflexivity on scientific practices is embedded in blogs as a process.

Scott L. Montgomery noted that "the online world not only opens science to wider participation and to new expressive forms, but by doing so it shows that the boundaries once assumed to divide scientific endeavor, in its essence, from other domains such as politics, economics, and the wider contemporary culture, are largely mirages" (Montgomery, 2009). Instead of making an artificially reductionist choice between addressing a restricted public of peers to communicate as part of the process of knowledge production or to disseminate science to less-partisan audiences, in different venues each following a different set of rules, bloggers can express themselves and build a mixed audience that makes no prior assumption of who shall be interested, who shall read and who shall respond. It breaks down many walls of the academic ivory tower and renders "the flow of ideas that circulate in an institution or disciplinary sector, more publicly available"[28], for anyone to see, benefit from, or challenge. In science just like in every other academic field, this can make a difference: for instance Professor Douglas A. Berman estimates that only about a half-dozen of his more than 50 law review articles and commentaries have ever been cited in judicial opinions, whereas his popular blog has been cited in more than a dozen legal cases, "including a dissenting opinion in a 2005 landmark decision by the U.S. Supreme Court"[29].

Science in the making gives often an unpredictable, or unanticipated perspective on science practice and science theory, revealing their natural messiness and flaws and reminding us that by definition they can never be perfect or absolute. Science blogs allow us to peep through laboratory doors and realize that the progress of research is more important to our daily lives than the "cold science" that has been canonized by the institution. As nanotechnologies find themselves at the heart of social controversies, the issue of climate change is being reframed by stakeholders, and stem cell research questions our ethical guiding principles, science can no longer wait for knowledge to be stabilized before making its voice heard. In risk societies transformed by complex issues, there is little hope in keeping the science in a safe place – while researchers stand at the front of heated debates and tireless negotiation work. Blogs maintain no such illusion.

The personalization of science blogging and the first hand accounts of research that it provides also help shift the interest from the content to the meta-discourse around it: who says what, in which context, based on which assumptions… This has traditionally been ignored in science communication, in spite of sociological work that has shown how much "backstage information" is critical to fully apprehend a scientific debate (Martin, 2000). Similarly, the discussion pages in Wikipedia makes visible the scaffolding of such intellectual edifice as an encyclopedic article.

Interestingly enough, the dialogic form that characterizes science blogs is common to both research and (modern) popularization of science. Scientific peers routinely engage

---

28 http://www.lukegilman.com/blawg/2006/07/31/blogging-in-higher-education-the-invisible-college

29 *Idem*

with each other to build intersubjectivity, while recent moves to restore public trust in science since the mid-1990s have come in the form of national public dialogue initiatives, like the "GM Debate"[30] or "Nanodialogues"[31] in the UK. Therefore, I believe that science blogs constitute the ideal candidate for a new communication channel, that naturally brings the scientists and researchers closer to their diverse audience, while inducing the public to participate more actively. Decades after sociologists of science convincingly demonstrated that formal communication and popular science represent the two ends of a continuum rather than a radical dichotomy (Shinn, 1985), it is more than high time to take action and explore new avenues that shove aside the usual categories.

Some (success) stories

Are success stories necessary to show the benefit of science blogging? There is a strong demand for it, even though it contradicts most of the central purpose of blogging. Blogging means building a long-term relationship with the readers, developing a shared understanding of what it is like to work in science or caring about science in the 21st century. In all but some high-profile cases, blogging is about small things, almost intimate. As Jill Walker puts it (Walker, 2006): "Blogs [...] are inherently social. Whether you have five readers or five hundred doesn't really matter, it's the knowledge that this will be read that is important. For a very new scholar, used to a world where three people will read a paper, grade it, and then put it away forever, that is an exhilarating and frightening idea."

When I started blogging as an MSc student in social studies of science in 2006, one of my goals was to show the readers what this field is like, what its stakes are, what it is trying to achieve. Over several hundred of posts, I delved into scientific frauds and ethics, the mechanisms of peer-review and publication, the relations between scientific expertise and democracy, the politics of technology, the robustness of scientific claims[32]... Nothing spectacular, but rather my individual attempt to shed some lights and build a conversation on a modest scale. During the protests that marked the academic year 2008-2009 in France, when academics tried to resist some political decisions that they thought were detrimental to the quality of their research and teaching, this community realized how its voice was hardly heard. They could not get their points across mostly because the stakes were complex and because they were victims of the "ivory tower" syndrome. This led to political discussions on how academic bloggers, by slowly changing the way they are seen and inviting more and more lay reader to enter their world, could change this situation[33].

The measure of success on science blogs consists first of all in the blogger's satisfaction with their writing and their use of the tool. Then comes the fact that their words are being linked-to, commented on or marked as "liked" on various social platforms – a metric of impact that might be compared with citation indexing in traditional science evaluation. It is also stimulating and rewarding just to observe the medium expand and witness so many clever discussions develop. For instance, French blogger Tom Roud

30 http://www.food.gov.uk/gmdebate/?view=GM+Microsite
31 http://www.demos.co.uk/publications/nanodialogues
32 http://www.enroweb.com/blogsciences
33 See http://www.enroweb.com/blogsciences/index.php?2009/06/02/398

(a pseudonym) offered some thoughts in March 2009 about the invention of agriculture and the subsequent evolution of modern societies[34]. Over the course of one month, many readers including an anthropologist, a chemist, a computer scientist and the archeology editor of a popular science magazine joined the conversation and contributed ideas to what turned out to be an open question i.e. just posed for the sake of the discussion and the pleasure of the argumentation. When this primary motivation is at the heart of a professional context or exercised in conjunction with one's own research interests, it can lead to advances in the accepted canon of knowledge or to setting the scientific record straight[35].

I shall exemplify one success story on the use of blogging in the context of scientific research. In January 2009, mathematician Tim Gowers started a massively collaborative mathematical project called Polymath. Given a problem that needs solving, the idea was to engage a large number of collaborators, spending each a minimum amount of time but reaching a smarter and faster conclusion thanks to the distributed diversity of creativity, skills and knowledge. After finding a problem that was amenable to this type of work, Gowers had to build a community (the readers of his blog were a good start) and find a way to make it work in an online, collaborative and transparent fashion[36]. The project ran for six weeks and was so successful that the online mathematical community has launched three other Polymath problems at the time of writing[37]. Interestingly enough, the projects make use of a blog and a wiki, i.e. a user-friendly website that anyone can edit like Wikipedia. Whereas the wiki is good at organizing knowledge, the blog allowed to keep a record of what has been done and who has done it. As one participant, who defines himself as "an amateur with just an undergraduate degree in math"[38] noted: "The linear format of the blog was wonderful in that it enabled my inspiration to come via observing the process of research. And the wiki was great too, since over the past two weeks I've been catching up on all of the pre-requisite math to actually understand what's going on. [A major progress of the project in the form of a mathematical proof was] a great moment for me because it solidified my whole understanding of the project – afterwards I went back and read through many of the comments again to better interpret the discussion. So to me, the real value of the project was simply that intermediate steps were being written up [on the blog], rather than the usual 'write up only when you have a paper to publish.'"

Will all scientists blog?

After stating the many advantages of science blogging, one may wonder whether this new communication medium will turn out to be the email-equivalent of the 21st century in research labs. In other words, might we see more and more scientists acknowledging "blog readers for comment and discussion" at the end of their academic publications? And prediction aside, would this be desirable?

---

34 http://tomroud.com/2009/03/26/agriculture-economie-et-evolution

35 http://www.rsc.org/chemistryworld/News/2009/July/27070901.asp

36 http://scienceblogs.com/christinaslisrant/2009/08/an_overview_of_the_polymath_pr.php

37 http://polymathprojects.org

38 http://gowers.wordpress.com/2009/03/10/polymath1-and-open-collaborative-mathematics/#comment-2710

One biologist has offered a good summary of the situation on his blog[39]. For him, scientists fall into two categories (although in reality there is of course a continuum between these): either they believe that "when you close a door you leave more outside that you protect inside", or they would prefer to work in their very own niche for fear of being plagiarized or scooped. Because of this fundamental difference, the former category will naturally be more attracted to blogging and doing science 2.0 than the latter. However, as they jump on the blogging bandwagon, the influence of research institutions could weigh in. More and more of them now promote science blogging and offer official tools for their researchers to blog, e.g. the Oxford Internet Institute[40] or the Ecole polytechnique fédérale de Lausanne[41] (Batts, 2008). With such endorsement from the hierarchy, there is little doubt in my mind that reluctant researchers will eventually start joining the conversation. Roger Pielke, Jr., blogger at Prometheus[42], a science policy blog affiliated with the University of Colorado Center for Science and Technology Policy Research, believes for example that "grounding the blog in an academic institution improves its legitimacy and reach" (Batts, 2008).

However, this trend raises many issues. First, once institutionalized, blogs need to be quality-controlled and their success gauged (Batts, 2008) – with the fear that blog metrics become another avatar of research metrics, and "metrics 2.0" based on audience, impact… end up suffering from the same biases attributed to the much criticized impact factor and citation counts. Second, blogging is fundamentally a bottom-up practice and a constant testimony of the autonomy of the research community, made of men and women rather than institutions and governments. As blogger and historian of science Thomas Söderqvist noted, borrowing categories used in political science, science blogs that are written from a first-person singular perspective represent the multitude whereas institutional science communication with a dominant perspective represents the Empire[43]. Therefore, the influence of power dynamics and conflict patterns within the research world are at the core of debate about the future of science blogging and should not be easily dismissed[44]. In fact, rather than institutional incentives, researchers might be more amenable to communitarian incentives. The short history of science blogging has shown that communities such as ScienceBlogs[45] in English or C@fé des sciences[46] in French have thrived by offering a good blend of science blogs and increasing the visibility of the whole, maintaining the momentum and allowing transient synergies between individual skills and centers of interest. In the case of the C@fé des sciences, it encouraged some members to blog outside of their *comfort zone* and dare to confront new topics – knowing that a colleague could add to the discussion or correct mistakes – as well as providing a framework in which to

39 http://www.recherche-en-cours.org/REC/Blogcast/Entrées/2008/9/12_Science_2.0_-_2_3.html
40 http://people.oii.ox.ac.uk
41 http://blogs.epfl.ch
42 http://sciencepolicy.colorado.edu/prometheus
43 http://www.corporeality.net/museion/2008/07/31/science-blogging-science-communication-and-the-multitude
44 http://www.corporeality.net/museion/2008/09/24/science-blogging-vs-institutionally-based-science-communication
45 http://scienceblogs.com
46 http://www.cafe-sciences.org

write joint or collective blog posts. The obvious success of bottom-up science blogs communities and their appeal to new bloggers are a critical factor in eliciting vocations.

A middle way between the marginalization and the institutionalization of science blogging is described by Shalley A. Batts *et al.* (Batts, 2008): "if groups of bloggers were to create their own initiatives and then seek institutional recognition, they might be able to engage in conversations about science on their own terms while continually proving to the institution – as they already strive to prove to their readers and peers – that the conversations they are engaging in are worthwhile. […] By initiating frank and open-minded conversations about shared goals, blogs and institutions can work together to advance the quality and scope of the ongoing global conversation about science we all participate in and depend upon." Another path for normalization is the increasing use of the term "online research notebooks", that takes "science blogs" away from the image of adolescent, intimate or controversial writings that are associated with the beginnings of blogs. As in the case of the French blogging platform Hypothèses[47], under the shared responsibility of several universities and research institutes, such an approach helps give the tool a fresh image and make it institution-compatible whilst departing very little from the original idea.

Now, what about these researchers and scientists who will not succumb to science blogging in the future? Shall we observe a defining difference between the "have" and "have-not" of science blogs?[48] Probably not: the quality of the have-not's research should be unchanged and it is safe to say that there will always be good researchers who do not blog. However, the difference will be for the society at large, who could lose out on the *positive externalities* of science blogging if its practice does not become widespread among researchers and scientists. In economics, this notion denotes the impact of a transaction on a party that is not directly involved. For example, fire-proofing a home also improves the fire safety of neighbors. We can hypothesize that some if not most of the value of science blogs lies in the externalities: they do not only benefit the researcher who can unleash his or her passion (Wakeford, 2008) but also other scientists and the general public. Values such as openness, access and engagement that matter to the society as a whole are well put to action in the science blogosphere… and in my opinion the world would surely be a better place if all researchers blogged.

Conclusion

Even if science blogging does not become the email of the 21st century, it is here to stay: I believe that what we make a big deal of today shall become a mainstream practice in the next decade. This will have a price: blogs epitomize freedom of speech and come with their share of harsh controversies, half-baked opinions and lazy reporting. They are also very effective rumor mills, what Matthew Chalmers exemplifies in the story of Tomasso Dorigo (Chalmers, 2009). A particle physicist and blogger, Dorigo commented in January 2007 on a small "bump" in data taken at the US particle physics laboratory Fermilab, which he

47 http://hypotheses.org
48 http://wiser-u.net/blog/2009/06/17/would-ramanujan-have-a-blog

suggested could be evidence of the much sought-after Higgs particle[49]. This led to a two-pages story in the popular science magazine *New Scientist* entitled "Glimpses of the God particle", which was soon picked up by mainstream publications such as *The Economist* and *Wired*. As the news spread in the blogosphere, the work that Dorigo exposed had yet to be reviewed by the 500-strong experimental collaboration of which he was part. He tried to distance himself from the story by claiming on his blog that the lack of statistical significance of the bump was not properly conveyed by the *New Scientist* journalist[50]. Dorigo later came under heavy fire from his collaborators, who have a strict system in place for releasing results to the wider world.

The blogosphere is also credited with a bestiary of readers and commenters that blogs users have to deal with[51]: the flamer who posts deliberately hostile and insulting *ad hominem* messages; the troll who posts irrelevant, controversial or off-topic comments; the kook who continually posts messages with no apparent grounding in reality; the shill who posts comments as a front for an unseen group or organization, usually at odds to the topics being discussed.

As a matter of fact, we are already learning to cope with these drawbacks and adapt to them. They are part of a bigger trend that is forcing us to rethink the current practices of research, which are as subject to the fundamental forces of change now as they have been over millennia. Be it in the way that priorities are established or collaborations arise, science will live in its century, as it always did. One day our offspring may laugh incredulously upon learning that in its early days blogging was often seen as a dangerous practice, a waste of time or, at best, a hobby – but we are not there yet. Graduate students or early career researchers are sometimes told by advisors and supervisors to stop blogging and concentrate on "real" work. This is also a reason why many science bloggers are writing under a pseudonym, which has the advantages of anonymity (you cannot be traced back to your real identity) without some of its inconvenience (i.e. with a pseudonym identity you can still build a clear reputation and track record on the web)[52]. However, this is another disruptive practice in the traditional scientific world, which generally uses an author's credentials to judge the merits of their work.

I hope that this essay has shed some positive light on science blogs and will motivate more scientists, academics, educators and citizens of the world to explore the blogosphere. One easy way to try out blogging is to do it for oneself. As a personal knowledge management tool, with timestamped entries, it can help one cope with the feeling of being subjected to excess of information and catch fleeting thoughts for the world to see. A few basic rules can help you make a good start[53]: choose a reputable blog host, keep privacy at heart, try to post twice a week and link to the material that you use. Inevitably someone else will come across your blog and approve, reply or comment. You will find yourself engaged in mind-expanding conversations in no time, on the very topics that matter to you. And along the way, you might find out that you are enjoying it.

---

49 http://dorigo.wordpress.com/2007/01/19/a-21-sigma-eccess-of-mssm-higgs/

50 http://dorigo.wordpress.com/2007/03/01/the-mssm-higgs-signal-buried-in-my-plot

51 http://www.searchlores.org/trolls.htm

52 http://reassignedtime.blogspot.com/2008/04/pseudonymity-is-not-anonymity-duh.html

53 http://www.scidev.net/en/practical-guides/how-to-set-up-a-science-blog.html

Acknowledgments

I thank Christina C. Pikas and Moira Cockell for their valuable input on the ideas and writing of the manuscript, as well as François Taddéi and Livio Riboli-Sasco for comments on the "Why blog?" survey. I am grateful to the C@fé des sciences community for the constant motivation in blogging about science and the many fruitful discussions around it. All responsibility for the opinions expressed in this paper as well as for any errors that may remain is of course mine alone.

Reference

Batts, S. A, Anthis, N. J. and Smith, T. C. (2008) "Advancing science through conversations: Bridging the gap between blogs and the academy", *PLoS Biology* 6, e240.

Borgman, C. L. (2007) *Scholarship in the digital age*, Cambridge, The MIT Press.

Chalmers, M. (2009) "Communicating physics in the information age", p. 67-80 in *Practising science communication in the information age: Theorising professional practices*, R. Holliman, J. Thomas, S. Smidt, E. Scanlon, E. Whitelegg (Eds), Oxford: Oxford University Press.

Claessens, M. (2009) *Science et communication pour le meilleur ou pour le pire?*, Paris, Quae.

Goldstein, A. M. (2009) "Blogging evolution", *Evolution: Education and Outreach* 2, p. 548-559.

Gregg, M. (2006) "Feeling ordinary: Blogging as a conversational scholarship", *Continuum. Journal of Media & Cultural Studies* 20, p. 147-160.

Martin, B. (2000) "Behind the scenes of scientific debating", *Social Epistemology* 14, p. 201-209.

Montgomery, S. L. (2009) "Science and the online world: Realities and issues for discussion", p. 83-97 in *Practising science communication in the information age: Theorising professional practices*, R. Holliman, J. Thomas, S. Smidt, E. Scanlon, E. Whitelegg (Eds), Oxford: Oxford University Press.

Patil, C. and Siegel, V. (2009a) "This revolution will be digitized: Online tools for radical collaboration", *Disease Models & Mechanisms* 2, p. 201-205.

Patil, C. and Siegel, V. (2009b) "Drinking from the firehose of scientific publishing", *Disease Models & Mechanisms* 2, p. 100-102.

Shinn, T. and Whitley, R. (Eds) (1985) *Expository science: Forms and functions of popularisation*, Dordrecht: Reidel Publishing Co.

Wakeford, N. and Cohen, K. (2008) "Fieldnotes in public: Blogs for research", p. 307-326 in *The SAGE handbook of online research methods*, N. Fielding, R.M. Lee, G. Blank (Eds), London, SAGE.

Walker, J. (2006) "Blogging from inside the ivory tower", p. 127-138 in *Uses of blogs,* A. Bruns and J. Jacobs (Eds), New York, Peter Lang.

3.5 Understanding the Science of Stories

by Noreen Golfman[1]

Once upon a time we understood the world by telling stories about it. The world was made intelligible through our shaping it into meaning. We relied on enduring elements of narrative to do so: a temporal continuity of events and characters experiencing those events in time. The meaning of narrative is always derived from the sequence of events themselves. Narrating particular elements in a sequence reveals something about the significance of those events. So it is that to tell a story is always explicitly to chronicle what happened and implicitly to evaluate or assess the events being recounted. One can always find value and value judgments in any narrative.

Once upon a time stories – let us call them narratives here, as well – made sense of nonsense, of the unknown or the unintelligible. Stories also made sense of experience, individual and collective. Stories gave us the world as truth. The emergence of ancient classical civilization comes to us in the form of Homer's hexameters, orally transmitted for centuries before being cast in written form; the birth of Indian religion and culture is carried forward through the epic Sanskrit verses of the *Ramayana* and the *Mahabharata*; the *Epic of Gilgamesh* tells us about early Mesopotamia; and to name only a few more ancient stories, we have fortunately inherited the *Aeneid*; the Old and New Testaments; *Beowulf*; the *Song of Roland;* and so on. Epic stories such as these are enormous windows onto the past, often the only views we have of the remote worlds in which the stories are set.

Centuries later, in the here and now, have stories lost their power to speak truth to us?

---

1 Noreen Golfman is Dean of Graduate studies and Professor of English Language and Literature at Memorial University Newfoundland.

Once upon a time

Shakespeare, perhaps most enduringly, promoted storytelling as the very glue of civil society. At the end of *Hamlet*, with a number of royal corpses spread across the stage and young Fortinbras approaching to sweep up Elsinore and restore order to a rotten Denmark, the dying tragic hero directs his only friend and loyal confidante, Horatio, how to go forward:

If thou didst ever hold me in thy heart
Absent thee from felicity awhile,
And in this harsh world draw thy breath in pain,
To tell my story.

(Act V, sc. ii, 284-87)

After Horatio promises to follow up, Hamlet repeats his instruction, drawing his last breath:

On Fortinbras: he has my dying voice;
So tell him, with the occurrents, more and less,
Which have solicited.

(Act V, sc. ii, 295-97)

Hamlet's "voice," although now silenced, will nonetheless be carried forward through his successor, a fresh prince and a new leader who will, ostensibly, learn the lessons of such tragic history. Horatio is the first designated bearer of that story, the trusted, neutral friend who will set the details right, and situate Hamlet in his rightful place in the history of Denmark, as much a victim as an agent of tragedy. For Shakespeare, modern civilization depends on knowing and repeating the story of its own constitutive elements. It could be said that all of Shakespeare's tragedies, including the History plays, convey this theme. Although humanity seems intent on repeating the same errors of judgment, and circumstances conspire to entrap even the most noble among us in webs of deceit, hope always resides in the possibility of learning from the past. Without a full story – or a proper history – of events, all that transpires would be meaningless. Repetition of that story not only reinforces but also informs civilization itself. Readers have not been poring over Shakespeare's works for over four hundred years because they promote futility or meaninglessness.

It is worth considering the critical life-affirming importance of story in Shakespeare against the virtual and deliberate absence of it in, say, the twentieth-century works of Samuel Beckett, where, as most famously in *Waiting for Godot*, nothing ever happens. Arguably, something does happen in *Waiting for Godot*, and decades of interpretive approaches underscore our urge to find meaning in its fragments of dialogue and thinly sketched characters. Nonetheless, few would disagree that the deliberate *avoidance* of story, of a coherent narrative apparatus into which characters can position themselves, generates a profound anxiety about the absence of meaning itself. While Shakespeare's tragedies insist on the socially redemptive possibilities of storytelling, Beckett's works question and debate the possibility of story altogether. From the Bard's point of view, and with apologies to

George Santayana, those who cannot remember the past are always condemned to repeat it. From Beckett's point of view, we have lost our memory entirely, and consequently our ability to tell each other our stories. Without such possibility, we are wandering, or waiting, in the world of the absurd.

This recurring trope of modern thought, that storytelling is a foundational element of culture and society, is, in the twenty-first century, under siege. We not only no longer look to stories to explain the world, but we often fail to recognize the inherent narrative properties of those discourses that claim they do. Truth has been evacuated from story and carried over instead to the domain of science and its progeny, technology.

This is a harsh statement but there is a lot of evidence to suggest that despite our generally warm academic embrace of inter and transdisciplinarity, the divide between science and the *human* sciences, as the French wisely deem them, or the humanities, as we more commonly say in English, is wider than it was when C.P Snow famously identified such a rift of "cultures" in the late nineteen-fifties. The further we move, if we are, indeed, so doing, from constructivist to scientific paradigms of knowledge, the deeper the rift becomes. Despite Stephen Jay Gould's eloquent disavowal in *The Hedgehog, the Fox, and the Magister's Pox* of C.P. Snow's (false) dichotomy (2003), and E.O. Wilson's earlier and ongoing advocacy of "consilience" (1998), and the importance of "the unity of knowledge," it is far from evident in the early part of this century, at least, that we are able to sustain an intelligent conversation about and between science and the humanities.

Who is telling the story?

This essay advocates the urgent need for that conversation. It takes as its jumping off point the time-honoured fact of and the inclination for story and its telling. Appreciating both our persistent need and ability to tell stories as a way to make sense of the universe has been taken up by at least one field of science – cognitive psychology. It is a commonplace that in the hierarchy of the sciences, psychology falls somewhere between biology and the "softer" field of sociology, with physics maintaining its position at the top of the heap. In this view, cognitive psychology has at least the merit of harnessing the hard blue-chip science of physics – particularly the physics of brain function – to its more dubious, softer subject of study, human behavior.[2]

It is impossible here to capture the scope and complexity of current, if still limited, work in cognitive psychology on the function and effects of narrative and narrative theory. But it is worth broadcasting at least the most cogent aspects of this research to indicate its potential. In particular, the subject of such research brings together our most profound questions regarding what we know to be true and *how* we know what we know to be true. And it is in the encounter between these two questions that our capacity for narrative structure emerges as a key element.

2 In this context it is well worth reading "Drawing Things Together," by French sociologist Bruno Latour, in which the author forcefully argues that the presence of graphs in scientific argument determines the degree to which a science is considered hard or soft or somewhere in between.

In 1991, an American psychologist published an article in a journal normally considered the preserve of humanities scholars. Admittedly, in "The Narrative Construction of Reality" psychologist Jerome Bruner is taking up a field of inquiry that literary theorists had already been exploring for at least a decade, and to some extent psychologists and cultural anthropologists had been catching up to, as well. Nonetheless, Bruner also acknowledges how much we had to date neglected to study the mind's ability to construct its own reality.

> It is curious how little effort has gone into discovering how humans come to construct the social world and the things that transpire therein… So while we have learned a great deal indeed about how we come eventually to construct and "explain" a world of nature in terms of causes, probabilities, space-time manifolds, and so on, we know altogether too little about how we go about constructing and representing the rich and messy domain of human interaction. (4)

Bruner goes on to explicate ten different aspect of narrative, openly challenging earlier rationalist or empiricist, thinkers, such as the iconic Piaget, who dominated the field for decades, noting that

> there is compelling evidence to indicate that narrative comprehension is among the earliest powers of mind to appear in the young child and among the most widely used forms of organizing human experience. (9)

These observations fall under the section titled "Hermeneutic composability" in the essay, a way of accounting for the human capacity to process knowledge by *constructing* and telling a story. Bruner is proposing that far more attention needs to be paid to the ways processing actually occurs, especially since it happens so early and so naturally. While it is true that Bruner's work speaks directly to the relation between the mind and the "messy" social reality the mind at once composes and processes, it is a small step from that social reality to considering the material or physical reality of the universe. What kinds of narratives and what sorts of narrative properties are we harnessing to tell the story of that equally messy universe? Not surprisingly, Bruner soon raises these questions himself.

What is especially appealing about Bruner's work in this context is that he is arguing against the current, dominant trend to treat the mind as an informational processor, as a computational computer capable of building computational computers, and back to a future of appreciating the mind as a creator of meanings. More recently, Bruner published a highly influential collection of essays in which he speaks directly to and provocatively about the construction of scientific discourse:

> The process of science making is narrative. It consists of spinning hypotheses about nature, testing them, correcting the hypotheses, and getting one's head straight. En route to producing testable hypotheses, we play with ideas, try to create anomalies, try to find neat puzzle forms that we can apply to intractable troubles so that they can be turned into soluble problems, figure out tricks for

> getting around morasses. The history of science… can be dramatically recounted as a set of almost heroic narratives in problem solving. (*Culture* 126)

Bruner is not the first person to be questioning the epistemology of science itself, but he is doing so *as* a scientist who has directed his career's work to the investigation of how we produce meaning. One need only consider the enormous influence of Thomas S. Kuhn on the way we think about the construction of scientific paradigms, a noun Kuhn borrowed from social science discourse, particularly linguistics, and entrenched in the philosophy of science.

What remains at least partly mysterious to many humanities scholars is the degree to which Bruner's provocations have been ignored or indifferently dismissed. Perhaps there is too much at stake, too many claims to truth to allow interrogation of the way those claims came to be. But, to echo Bruner, if we stopped seeing narratives as fictional representations of life but as systematic formal construction of life we might come up with at least a different story for what we are—and in turn what the world is. To be sure, his call has not gone completely unheeded, and some scholars have taken up the challenge of identifying the mechanics of such "heroic narratives," otherwise commonly known as *discoveries*.

Notably, of all people an English professor has torn into the vexing and often marginalized subject of the relation between narrative and consciousness with gusto. David Herman continues to publish extensively in his field, pushing beyond traditional disciplinary boundaries to create new space for interdisciplinary scholarship. A reviewer of Herman's edited collection, *Narrative Theory and the Cognitive Sciences*, writes in the journal *Style* that

> Despite the astonishing number of books and articles with the words "consciousness" and "cognition" in their titles published within the last decade, there remains a significant and disappointing disconnection between the "scientific" fields of cognitive research and literary criticism and theory.

Note the qualifying inverted commas around the word scientific, as if the reviewer needed to protect himself from charges of too comfortable a purchase of his own statement. Herman has, indeed, been unabashedly progressive as a forceful, single engine in the field, promoting as the title of a landmark article describes it "Narratology as a cognitive science." Noting in 2000 that the impressive 1999 publication of the *MIT Encyclopedia of the Cognitive Sciences* lacks an entry for Narrative or Story, Herman advocates narrative theory as a subdomain of the cognitive sciences, an invitation not answered to date. "Both language generally and narrative, specifically," Herman argues, can be viewed as tool-systems for building mental models of the world" (2). Ambitiously, he strives to bring into dynamic play current scholarship in anthropology, linguistics, computational intelligence, philosophy of mind, and cognitive psychology. By way of rich example, he mines the *MIT Encyclopedia* to show how entries in areas such as "Problem Solving," "Cognitive Maps," and "Schema," among many others, already tease out narratological pathways, such as "selecting actions that will achieve goals," "dynamic and static knowledge representations – that is, stereotyped sequence of events … and [descriptions of] states of affairs or situations" (4). Further he invokes Bruner's model of consciousness based on two main strategies for making sense of the world: one being "paradigmatic or logico-classifactory thinking" and the other, as previously mentioned, narrative thinking (5). Sometimes, these two modes overlap or coalesce. As Herman says, "Bruner's work suggests that stories should be examined as a fundamental sociointeractional resource for the development and refinement of human intelligence" (5).

More pertinent to the overall aim of this essay is Herman's unpacking of the *MIT Encyclopedia* introductory piece on "Computational Intelligence" (shrewdly not defined as Artificial Intelligence), in which the authors explore the program "architecture" necessary for establishing computational capabilities. Everything depends on what kinds of tasks might be demanded of such intelligence, but Herman persists in pointing out that essential characteristics of narrative would be required for such functions as "Speech Synthesis," "Natural Language Generation," "Processing," "Causal Reasoning," and "Planning" (6), to name only a few. These tasks necessarily depend on sequences of events or words, on chains of causes and effects, on strategic deployment of signifiers both added and omitted, obvious features of narrative structure. "In short," Herman summarizes, "the demands placed on a system with genuine narrative intelligence will be considerable" (6).

This is but a brief and oversimplified version of what Herman takes up in his article, which is well worth reading to engage in the possibilities of truly rich interdisciplinary research. As he works methodically though each of the major sections of the *MIT Encyclopedia* he pushes us further along towards a synergy of narrative and scientific thinking, boasting nothing less than that "narrative theorists are in a position to help reframe the central questions" linking culture, cognition, and evolution. As he argues, for example, whether we are examining humans or animals at social play, we should be bringing to bear the attributes of narrative to inform our research, thereby advancing our understanding of a "richer repertoire of play-indexing cures, than those hitherto recognized" (9).

In this vein, it is not surprising that Professor of Neurology & Psychiatry at Columbia University, Oliver Sacks, is a best-selling science author, for he satisfies a hunger readers have for telling stories about stories themselves. Sacks believes firmly with Bruner that the human mind is designed to tell stories, and he deploys all the time-tested tricks of narrative to capture the most complex and mysterious of human conditions. Assessing the narrative strategies of Sack's work in a wonderful review essay entitled "The Ecology of Writing Science," William Howarth cites among others Lewis Thomas who maintains that words are more important to science than 'getting all the numbers right'" (104). As overstated as this may be, the sentiment underscores the value that narrative brings to our understanding of how the mind works, and how what the mind is itself working *on* might be structured.

> Standard neurological journals prefer a mechanical discourse: dry, storyless, governed by abstract nouns, tangled clauses, and passive verbs—the inert style of so-called "life sciences." Even direct statements such as "Parkinson's disease primarily involves degeneration of dopaminergic neurons in the human brain" describe disease as chemical actions rather than human events. Sacks instead uses anecdotal material for its local and eccentric properties, as specifics that assess the validity of suffering. (109)

As becomes obvious reading Sacks' work, he believes in using case-history narratives as the best way of illustrating illness and its treatment, and it is hard not to wish others followed his example. How much more accessible and therefore meaningful would the troubled or mysterious worlds appear to us.

Scientists who sing to us through a reliance on narrative, as well as parable and anecdote, are rarer than ever. Loren Eiseley, a iconoclastic "naturalist" of the last century, embraced, like Oliver Sacks, a personalizing approach to his subjects, in direct opposition to the dominant, impersonal discourse of modern science. But these figures are rare, and it could be argued they possess a gift of writing that carries them gracefully away from the anemic, stilted language of their peers. Above all, they appreciate the power of narrative as both a shaping instrument and a representation of the mind itself at work. If disciplinary boundaries are worth shedding they must do so through a new form of wring, as well, one that moves towards more, not less, of the human in the narratives of the universe, and thus more self consciousness about the potential of language.

A Popular Story

It is humbling to consider how many millions of people have been drawn to the big screen to see arguably the most popular story of the first decade of the century. James Cameron's *Avatar* is not only an experiment in a new cinematic technology but also an epic story of the not-too-distant future, one in which science is capable of creating both beautiful hybrid creatures and highly efficient machines, and wreaking terror on other alien worlds that happen to be in the way of the march of technology and resource exploitation. Creator Cameron famously created an entirely new language with its own vocabulary spoken by the idealized Na'vi, the peaceful occupants of Pandora who are confronted by menacing earthly invaders. While it is true that Avatar has drawn millions into theatres partly because of its breathtaking special effects, indeed, its marshalling of new technologies to generate a dazzling spectacle, even the most hardened critics have agreed that the story, old-fashioned in its simplicity and familiar in its broad-strokes treatment of character and conflict, is what drives momentum and sustains audience interest. However diverse or bizarre the interpretations to *Avatar* may be, the movie does prove the appeal of a story well told. Clocking in at well over two hours, *Avatar* also belies the cynics who argue that today's audiences can withstand no experience longer than it takes to send or receive a text message. It is not that there is any necessary virtue in length. The point is that, as noted critic Roger Ebert writes, but

> the film doesn't feel too long. It contains so much. The human stories. The Na'vi stories, for the Na'vi are also developed as individuals. The complexity of the planet, which harbors a global secret. The ultimate warfare, with Jake joining the resistance against his former comrades. Small graceful details like a floating creature that looks like a cross between a blowing dandelion seed and a drifting jellyfish, and embodies goodness. Or astonishing floating cloud-islands. I've complained that many recent films abandon story telling in their third acts and go for wall-to-wall action. Cameron essentially does that here, but has

invested well in establishing his characters so that it matters what they do in battle and how they do it. There are issues at stake greater than simply which side wins.

Some are arguing that in the early part of the twenty-first century there is no longer any belief in either the power of stories, or no stories with any insight worth attending to. Current debates about the future of the novel, for example, tend to have taken on notions about how ideas and information and the "reality" they convey should be presented, especially in an age where information is so routinely bite-sized, fragmented, random, chaotic. Consider, as some do, the disconcerting leap from the weighty expression of a *War and Peace* to the 140-character-bound limitations of a tweet.[3] But a story is a story is a story, and in some ways a tweet is a contemporary version of the lyric or haiku, a "found poem" in which meaning (as in "I am slowly moving my feet towards the end of the garden. Shall we meet for coffee at 7 or r u free?") easily shapes into a discernible narrative arc.

With all the decrying about the reduction of the personal to a tweet, or the proliferation of punctuation-challenged text and social networking entries, it is tempting here to identify the persistent and ubiquitous reliance on narrative that characterizes these new communication phenomena. Everyone still wants to tell a story. Some have been saying for some time, as with Paul Ricoeur and Judith Butler, to name two obvious icons of contemporary theory, that identity itself is a narrative construction, and the field of literary theory has been richly informed by studies of ourselves as beings who are interpreted and who are transformed through the process of interpreting others and their narratives.

Science is telling its own stories about our universe. It would be fruitful to recognize those stories *as* stories, and to move on from there.

References

Bruner, J. S. (1991) "A Narrative Construction of Reality," *Critical Inquiry* 18:1, p. 1-21.

Bruner, J. S. (1996) *The Culture of Education,* Cambridge, Mass., Harvard University Press.

Butler, J. (1999) *Gender Trouble: Feminism and the Subversion of Identity*, New York, Routledge.

Cameron, J., dir. (2009) *Avatar*, Twentieth-Century Fox.

Conley, T. (2005) "Review of David Herman, ed. Narrative Theory and the Cognitive Sciences," *Style,* September 22. http://findarticles.com/p/articles/mi_m2342/is_3_39/ai_n27869159/pg_4/

Ebert, R. (2009) "Review of *Avatar*," *Chicago Sun-Times.* http://rogerebert.suntimes.com/apps/pbcs.dll/article?AID=/20091211/REVIEWS/912119998/1023

Eisley, L. (1972) *The Unexpected Universe* [1969], New York, Harcourt.

Gould, S. J. (2003) *The Hedgehog, the Fox, and the Magister's Pox: Mending the Gap Between Science and the Humanities*, New York, Harmony Books.

Herman, D. (Ed) (2003) *Narrative Theory and the Cognitive Sciences.* Stanford, Center for the Study of Language and Information.

Herman, D. (2000) "Narratology as a Cognitive Science," *image and narrative* 1.1, p. 1-26. Available online. http://www.imageandnarrative.be/narratology/davidherman.htm

Howarth, W. (1990) "The Ecology of Writing Science," *Modern Language Studies*, 20.4, p. 103-120.

[3] See David Shields' *Reality Hunger: A Manifesto*, in which the author attempts to "reframe" how we think about art and truth.

LATOUR, B. (1990) "Drawing Things Together," p. 19-68, in *Representation in Scientific Practice,* LYNCH, M. and WOOLGAR, S. (Eds) Cambridge, MA, MIT Press.

RICOEUR, P. (1992) *Oneself as Another*, trans. Kathleen Blamey [1990], Chicago, University of Chicago Press.

SHIELDS, D. (2010) *Reality Hunger: A Manifesto*, New York, Knopf.

SNOW, C.P. (1993) *The Two Cultures*, Cambridge, Cambridge University Press.

WILSON, E.O. (1998) *Consilience: The Unity of Knowledge*, New York, Knopf.